WHAT THE BUILDER WON'T TELL YOU

THE ESSENTIAL HOMEOWNER'S GUIDE TO AN ENERGY EFFICIENT, HEALTHY HOME

JUDITH LEARY-JOYCE

Copyright © 2025 Judith Leary-Joyce

All rights reserved.

The right of Judith Leary-Joyce to be identified as the author of this work has been asserted by her in accordance with the Copyright Designs and Patents Act, 1988.

ISBN: 978-0-9930772-6-5

No part of this publication may be reproduced, stored in a retrieval system or transmitted, in any form or by any means, electronic, mechanical, photocopying, recording, or otherwise, without prior permission from the publishers.

This book is sold subject to the condition that it shall not, by the way of trade or otherwise, be lent, re-sold, hired out or otherwise circulated without the publisher's prior consent in any form of binding or cover other than that, in which it is published and without a similar condition including this condition being imposed on the subsequent purchaser.

Judith Leary-Joyce is not an architect, engineer or building expert. This book does not take the place of advice from professional architects, engineers, builders or specialist building suppliers. Every effort has been made to ensure that the content provided in this book is accurate and helpful for readers at time of publishing. However, this is not an exhaustive treatment of the subject. No liability is assumed for losses or damages due to the information provided. You are responsible for your own choices, actions, and results. You should consult your architect, engineer or building professional with specific questions about your renovation or retrofit project.

Produced and published in August 2025 by AoEC Press

For all the grandkids

Those already with us and all those to come

PRAISE FOR WHAT THE BUILDER WON'T TELL YOU

This book made me see renovation differently — sustainable design isn't a compromise, it's an upgrade. **Josie Gibson, TV Presenter***

Buildings are one of the biggest sources of carbon emissions - bad for the planet but also bad for wallets, so it's invaluable to have expert insights on how to improve our homes. This book demystifies the improvements available and gives excellent advice on how to work with tradespeople for efficiency and electrification. **Greg Jackson, CEO Octopus Energy**

A powerful guide to transforming our homes for a sustainable future — practical, inspiring, and essential reading. **Georgina Burnett, Home Coach and TV presenter***

This book makes retrofit feel not just possible, but exciting. Judith gives homeowners the confidence and inspiration to create warm, healthy homes that don't cost the earth, and actually help us protect it. **Sam Bentley, Sustainability Advocate ***

This is exactly the kind of plain-spoken, good-humoured guide we need right now — a book that helps ordinary homeowners ask the right questions and make real, climate-smart choices. Judith demystifies eco-renovation without ever preaching, and she does it with warmth and wit. **Dave Borlace, Climate communicator ***

This is the book we all so desperately need, homeowners, builders *and* planet! **Michelle Ogundehin Home Therapist***

Judith brings clarity, empathy, and hard-earned wisdom to one of the most confusing areas of home improvement, retrofit, and covers everything a homeowner needs to know today before undertaking any renovation. **Ellora Coupe, herownspace***

There's no denying it - we have an almighty task on our hands: retrofitting and future-proofing the UK's housing stock. So, thank goodness for people like Judith, who are supremely passionate, driven, and determined to untangle all the quirks of the current system, making it easier for the rest of us to follow with confidence. This guide is an essential read for anyone starting their clean-energy home journey. **Imogen Bhogal, Everything Electric Show, Green Queens ***

I'm at the very start of my eco renovation - (feeling equal parts excited and daunted) - and I'm determined to make every decision as sustainable as possible. I'll learn as I go, but Judith's experience in this book does a lot of the heavy lifting, answering the questions people like me face when aiming for a low impact build. This is a practical journey, not a perfect one and with Judith's guidance in this book - that comes from everything she has experienced herself - I feel more confident and informed with my choices. **Lizzie Carr MBE***

CONTENTS

Acknowledgements	ix
1. Why The Builder Won't Talk About Retrofit	1
2. Understanding An Energy-Efficient Home	13
3. Retrofit Benefits The Builder Doesn't Know About	36
4. How Construction Impacts The Climate	44
5. How To Manage The Builder	57
6. Conventional Materials The Builder Will Use	76
7. What You Need To Know About Costing	101
8. The Professionals That Can Help You	125
9. What You Need To Know About Contracts	143
10. What The Builder Won't Know About Retrofit	153
11. All You Need To Know About Natural Materials	183
12. The Challenge Of Sustainable Heating	210
13. How To Get The Renewable Energy You Want	234
14. What You Need To Understand About Glazing	255
15. Why Recyling Won't Occur To The Builder	267
16. Why Healthy Homes Aren't On The Builders Radar	282
17. Instagram Wisdom	293
Glossary	299
About the Author	305
Also by Judith Leary-Joyce	307

ACKNOWLEDGEMENTS

So many people have supported me through my learning and writing of this book.

Particular thanks to:

Ellora Coupe (Her Own Space*) for being a fellow building bore and constant support. Your endless knowledge and expertise, plus your willingness to answer sudden What'sApp message on a Sunday has been remarkable.

Suzanne Arnold* for your support, sensitive editing and making me giggle along the way. And to your Mum for agreeing to be in the book!

Brad C* Thanks for doing such a great job at short notice on the index and being so easy to work with.

Lee-May Lim* for creating my perfect book cover.

Valerie Couillard (The Place Between*) for your painstaking care over contracts and the professionals we work with.

Leah Robson, (Your Energy Your Way*) for your thorough, detailed reading of the chapters on renewable heating and sustainable energy.

Nikki Lambert (Lambert Homes Builds*) for keeping me on track with 'What the builder won't know about retrofit and How costing works.

Alison Phillips*, Amy Isted*, Amy Somerville and **Aniko Hegedus*** for teaching me something new in every conversation.

To my dear friend - you know who you are - for the endless care and support you give me. You help me stay focused on the work we need to do.

Judy Newton-Davis - for your unerring support and never ending confidence in me.

To my family who put up with my enthusiasm for hemp insulation and still encourage me to keep going.

Finally to **my husband John**, for supporting me, whatever wild scheme I come up with.

CHAPTER 1
WHY THE BUILDER WON'T TALK ABOUT RETROFIT

There's a revolution just beginning in house building and we need more people to know about it – including the builders.

It's the shift towards comfortable energy-efficient homes that are cheap to run and don't cost the earth. They need to be warm in winter and cool in summer. And be ready to withstand unexpected climate events.

I'm talking about retrofit

I've been talking about retrofit for the past four years because it's a game changer. It's taking a tired, cold, draughty house into a future of low energy bills (or even no energy bills), top air quality and comfort.

Everything we need to know is available – it's not rocket science; most people just don't know it's an option. We assume builders are the experts, so we hand over responsibility for our homes and hope for the best.

Truth is, you'll get what the builder wants to give you. Which may be a warm, healthy home that's a delight to live in. Or it might be a beautiful-looking home but riddled with condensation and mould that leaves you dreading your energy bill.

Retrofit is like having a baby

I've done both and there's a big similarity. Both are:

- hard work and demanding
- really painful on occasions
- so wonderful that you immediately forget how horrible it's been.

Until you begin again, of course.

There is also a similarity in how people talk to you about the process. There will always be some who can't wait to share their horror stories. They're determined to go through every gruesome moment with all the gory details. They forget to tell you how amazing it is once the pain is over.

And then there are others who just want you to see the wonder of the end result, without any reference to the struggle and chaos.

WHAT WE NEED IS BALANCED INFORMATION

That's what this book will give you. I won't back off the struggle – it's definitely there – but it's paving the way for something fantastic.

If we only share the best and the worst, we omit the valuable information that could help navigate the process more easily.

For example, it's useful to know that builders sometimes:

- need money before they even begin, to buy materials, and it doesn't mean they're going to cheat you

- forget this is your home and leave dirty cups and cigarette butts around the place
- don't believe that heat pumps work, so will point you back to a gas boiler
- don't even consider solar panels
- have no idea about natural materials, because they're just used to the synthetic option.

It's all manageable when you know what's likely to happen. You can plan ahead – or at least not be thrown when something suddenly changes. And when you know the terminology used, you can take your place in the conversation.

As homeowners, we are the front line

We have buying power and we need to use it. Now is the time to state clearly what we want: to make sure home improvement actually means improvement and not a step backwards.

To take your place in this major shift you need some idea of what's possible and you need information. The first step is to be more demanding – you deserve better than this. You have the right to a warm and healthy home, but there just aren't enough providers who understand. Which means it won't be offered to you on a plate – you'll have to ask, demand, require it.

Builders won't tell you this purely because they don't know themselves. Once we consistently demand retrofit and energy efficiency, they'll have to learn or lose our business. It's time for us to move the dial. In helping ourselves we'll help raise standards, so those who don't have the buying power will also benefit.

That's what this book is about. I want you to have the baseline information and the terminology you need, so you can be part of the change.

We can do this and make a real difference: for ourselves, tenants and future generations.

So let's get going.

It's for the grandkids

I've cared about the environment since the early 1970s. It's obvious to me that we need to care for the earth – it is our actual home. We can't survive without it, so why would we treat it so carelessly?

This is about our own safety and comfort, but more important than that – this is about the grandkids. They have to live on the earth we leave them and that's not looking safe or attractive right now.

I'm always looking for the next change I need to make. So when I saw the impact of our own retrofit, I got really excited. Here was the perfect climate action. Lower energy bills, greater comfort plus a house that's worth more – <u>and</u> you lower your carbon emissions by a shedload.

But we have a big job ahead of us. The building professions are not fully on board yet. They prefer to work as they've always done – using materials produced from oil and gas, paying little attention to energy efficiency and refusing to give proper consideration to ventilation.

We deserve better – and so does the planet. It's time to get our acts together. You don't need to be an expert. You just need to know enough to ask the right questions and get what you need.

It's a bit more effort – but good fun when you get into it. And the grandkids are worth it!

WHAT'S THE DIFFERENCE BETWEEN RENOVATION AND RETROFIT?

Renovation improves how a space looks and how well it suits the needs of the residents – new rooms, new kitchen/bathrooms, improving gardens, décor, new floors… All the elements that make an attractive home.

Retrofit transforms how a home performs in terms of cutting energy usage, reducing bills, making the home warm in winter and cool in summer. This change is achieved by putting into the house what wasn't included when it was built: insulation, airtightness, controlled ventilation, sustainable heating, renewable energy.

Given the climate issues we face nowadays, retrofit takes on an even bigger significance. We'll always want to live in a comfortable home that doesn't cost a fortune to run. Now we also need to future-proof – so that home is fit for whatever comes our way.

We need a home that isn't costing the earth – in every sense of that phrase.

Retrofit is a game changer

I was introduced to retrofit through my personal experience. Once I realised what a game changer it is, I was converted and I've been shouting about it ever since.

I've lived in a standard three-bedroom Victorian end-of-terrace with my husband John for nearly 50 years. Over those years we've added rooms, taken out walls, changed windows. Then came the most recent project – an extension to the back of the house in place of a much-loved but too hot/too cold conservatory.

John wanted to build a new room and knock through into the kitchen, giving us an open-plan sitting area, cooking and dining space. I didn't like the idea – open plan means wasted energy to me. Finally we found a compromise. We would hire an eco-architect to design the extension.

It was only once the work had begun that we realised how daft this was. What's the point of putting a well-crafted, airtight, insulated room on the back of an old leaky house? What were we thinking! The architect thought it would be 'good enough' for the air source heat pump we wanted, but who wants to settle for 'good enough' when perfect would be so much better?

After an evening of soul searching, we decided we'd make our home a legacy for the future. This would mean taking the principles used in the extension and applying them to the rest of the house. No small job – but heigh-ho – we'd started, so we were going to finish.

Imagine the builder's face

Imagine Pete walking in the next morning to find that his work had increased exponentially!

Pete works on his own, bringing in different trades when needed. He is a brickie by trade, but has expanded into building extensions. He works on one job at a time – which is what appealed to us – so we would be the sole focus of his attention until the job was done. And we knew we got on together well enough.

BUT: an eco builder he wasn't. He had his own way of working and generally stuck to it. The gold for us was that he was open to discussing options before agreeing next steps with us, even when it was unfamiliar to him.

SO WE BEGAN

We were all on a steep learning curve. With the architect no longer available, we followed the blueprint we had while I got onto the laptop and began exploring.

Where to begin? I tried searching for 'renovation' first, which led me to beautiful kitchens, bathrooms, carpets and curtains. It showed me what our new home could look like, but nothing that helped me now. And we were back to the brick, getting dustier by the minute.

The turning point came when I discovered the word 'retrofit'. Now the world opened up before me. I found information about breathability, sustainable insulation, ventilation and airtightness. I read sales sites galore and endless dense, erudite papers on the strengths and weaknesses of wood fibre - it was hard going - but I was finally on the right track.

Climate vigilante rides again

With hindsight and the knowledge I have now, I realise we were saved from disaster because I'm a climate vigilante.

So great is my concern that I refused to use any materials that were detrimental to the environment. The very thought of synthetic PIR insulation (silver-coated solid foam) made from oil brought me out in a rash.

THE TOUCHSTONE

After long discussions we agreed to make our home a legacy as well as comfortable to live in. To do that we needed to agree our priorities - our touchstone

For us the touchstone was, in order of priority:

- Climate first and foremost
- Cost
- Interior design.

Beautiful finishing touches are important, but they can be done at any time. There is just one opportunity to insulate and make airtight. It's not often we go back to the brick in a project, so when it happens it's the perfect moment for making a home warm and energy efficient.

Turns out the touchstone was perfect. We did a good thing for the planet by creating a home that needs 75% less energy. We were also more comfortable than we'd ever been in this old, leaky bucket of a house.

And that same touchstone saved us from our ignorance.

An old house like ours – built before 1940 – needs to be breathable. The standard rigid insulation used by builders – PIR – is not breathable. I had no idea why that mattered, but I did know PIR was made from petrochemicals so I was never going to use it. So we just continued with wood fibre as recommended by the architect for the extension.

I now know we'd stumbled on exactly the right thing for our house, courtesy of the touchstone. By refusing synthetic materials we won the battle over condensation (see chapter 11). Without knowing why, our love of the planet saved us. What a lesson!

Was it worth it?

Here's the good news and why I've been shouting from the rooftops ever since:

- Our energy usage reduced by 75%.
- Our EPC went from a D to a B.

- Our house is now worth at least £90K more than the equivalent house without a retrofit

AND we are comfortable and warm. My constant companion, the cashmere scarf, has been relegated to the back of the wardrobe and we no longer know what the weather is like outside the house.

So yes, it was worth it for our comfort and bank balance.

SO WHY AREN'T BUILDERS SHOUTING ABOUT RETROFIT?

Why don't builders talk about this? Maybe they think we won't be interested? Maybe they just don't know? Maybe they need a push to venture out of their comfort zone? Whatever the reason, it's a very rare builder who'll suggest retrofit or talk about natural materials.

Which brings me to the job before us

Remember that buying power? You're in a position to ask for what you want and stay engaged to make sure you get it.

But to do this well, you have to understand what's needed, the impact it'll make and the quality that'll deliver those results. You need to know what you don't know, so you're not shunted into just accepting what the builder wants.

And that's why I've written this book. I want you to join me in this retrofit revolution. And for you to be part of the next steps, you need to know what the builder can't or won't think to tell you. And, if they do tell you, you need to be able to understand what they're saying.

For the essentials and details of the retrofit process, read *Beginner's Guide to Eco Renovation*. This is the book we needed when doing our own house. It wasn't available then, so as soon as we were done and proven, I wrote it for you.

This book is a sister to *Beginner's Guide*. The aim is to put you in a position where you can work with the builder to get what you want in terms of energy efficiency and healthy home.

We will cover:

Understanding the basics

1. Why the builder won't talk about retrofit: why the builder won't – or can't - tell you what you need to know
2. Understanding an energy efficient home: how energy efficiency affects the way building work is done, what it looks like, why it matters.
3. Retrofit benefits the builder doesn't know about: the benefits of making home energy efficient, the standards that can help guide you, how to avoid renovators regret
4. How construction impacts the climate: harm done to the environment and how we can change that.

Working with the builders - what you need to know

5. How to manage the builder: how to find a builder, what questions to ask, what to tell them before they quote, what they won't think to say about what's going to happen.
6. Conventional materials the builder will use: what to expect from a building company, synthetic materials they automatically use and their implications, basic terminology / labels that will help you.
7. What you need to know about costing: different ways

that builders charge according to their working style, how money is handed over, who pays for what.
8. The professionals that can help you: the professionals you need before you get to the builder
9. What you need to know about contracts: what you need to know about contracts, insurance and safeguards.

Setting up to be energy efficient

10. What the builder won't know about retrofit: retrofit essentials: airtightness, insulation, ventilation, breathability,
11. All you need to know about natural materials: the importance of natural materials and their qualities
12. The challenge of sustainable heating — different options: air source, ground source, biomass etc. Problems with heat pumps: installation and homeowner mindset
13. How to get the renewable energy you want: different solar options, pro's and con's of batteries, selling to the grid.
14. What you need to understand about glazing: what you need to know about windows and doors. Their role in creating an energy efficient home.
15. Why recycling won't occur to the builder: making do and recycling, managing water consumption, water quality, reusing waste and reducing landfill.
16. Why healthy homes aren't on the builders radar: the elements of a healthy home and how it maps onto retrofit.
17. Instagram wisdom: what Instagram followers wish the builder had told them.

LINKS AND FURTHER INFORMATION

At the end of each chapter there is a QR code. This links directly to my website where each chapter will have its own page with links, pictures, blogs related to that specific subject.

Wherever you see an asterisk * that's your clue to go looking for more information and I've left you some space to write any notes you need.

Off we go!

To find the information indicated by the * follow the QR code.

NOTES

CHAPTER 2
UNDERSTANDING AN ENERGY-EFFICIENT HOME

With energy prices rising and the world becoming increasingly unpredictable, energy efficiency is one way we can feel more secure. But building professionals haven't caught on yet: cutting our energy bills just isn't high on their list – or even on their list.

Builders and building professionals won't:

- look at plans and suggest ways to reduce energy use
- talk about what your home might need in the future
- tell you about different materials you could use and the benefits they bring.

Generally, builders, architects, designers… will respond to what you want with what they have in their armoury. It's like that old saying: 'if you have a hammer, then everything is a nail.' If you want to have more space in your house, they'll have a set piece to offer. Bit like tried and tested recipes. It's clearly worked in the past – at least in terms of providing a livelihood for the building professionals and short-term gains for homeowners.

THE MISSING LINK

But there is a big piece missing. Our homes are not ready for the future. And our homes are not of good quality. We have the worst housing stock in Europe and any work done to improve that has been driven by short-term thinking.

Someone needs to put a boot up the backside of the profession and as homeowners/customers we may have to be the ones to do it.

Fair's fair

To be fair, there are some shining examples out there:

- Passive Houses built in a sustainable way that requires no energy input at all*
- Octopus Energy driving for zero bills homes*
- York Council building 400 zero carbon homes
- Barratt Homes planning to build net zero homes from 2030
- the New Homes Net Zero Transition Plan being created to provide a pathway and metrics for decarbonisation*.

All this is brilliant – and there are many more than I've mentioned here.

But we're still at the stage where each achievement is something to shout about. It's time zero carbon and energy efficiency were commonplace and of no particular interest. Some in the profession are working really hard on this; some are just carrying on as if nothing is happening

Do builders need support?

I sometimes wonder about builders and what life is like for them. They're doing the job they've always done – in the same way, with the same materials, and the same people. Do they ever get fed up and wonder how life could be different?

Some will always want to hold onto the 'same old, same old' but there will also be some who'd love to upskill and find new ways of working. Not to mention concerns for their health and wellbeing. I recently heard of a carpenter who expressed concern about 'the amount of PIR dust I've breathed in through the last 30yrs without realising'. (See chapter 6.)

Support for the building profession needs to come from higher up, but until that happens, we can help move the dial by asking for materials that are good for the builder, our homes and the planet.

Homeowners as change makers

As homeowners we can join the vanguard and use our buying power to speed up the process. The more of us who ask for energy efficiency, enquire about future-proofing and the materials that will get us there, and make clear that we want to be sustainable, the sooner builders will realise this is serious.

Money is a great driver of change. So our clout comes from turning down the old-style 'get it done, get paid and move on' builders and making clear we'll only employ those who understand energy efficiency. Then we'll start to see change.

WHAT MAKES A HOME ENERGY EFFICIENT?

To do all that, we need to understand what will cut energy bills and ensure we're really comfortable in our homes. Once we have that clear in our minds, then we'll know what to ask for.

At present, home improvement focuses on making the place beautiful – colours, carpets, kitchens, bathrooms, smart floors… The list goes on. The end result is lovely to look at and something we can proudly invite friends into.

What we miss is the way the home functions. We'll talk about the sofa and curtains for hours, but it's rare for insulation to be the subject of dinner party conversation.

Until now. If we're taking our place in the energy efficiency revolution, we need to talk and get excited about insulation – that's how we get the word out.

Mum, you've become a building bore

When my daughter Miriam told me I was a building bore, I was taken aback. I'd not long finished writing **Beginner's Guide to Eco Renovation*,** *so my head had been embedded in wood fibre, lime plaster and heat recovery units for months. Clearly not where my kids and grandkids were focused!*

We'd started out to build an extension. Retrofit was just an add-on. With standard builders who had no idea about retrofit and us being no better, I had to start learning. Fortunately – for me, anyway – I loved it. No one could be more surprised, but thank heaven, otherwise we'd have been in a terrible mess.

If you've ever written a book you know it requires total immersion, so my mind was running on a single track. My poor family! I was forever waxing lyrical about heat pumps when they were ready for a Friday night drink and cashew nut.

Then I realised she was right – I was a building bore! So I decided to embrace it.

Four years on and I'm still going strong. If anything, I'm even more enthusiastic – there's so much to learn. Just don't get me onto clay plaster if you're in a hurry or have a train to catch. I do my best to read the room, but I can't promise. Waxing lyrical has become second nature!

WHAT WE NEED TO KNOW

'Retrofit' is the word that makes all the difference, but few people have heard of it – unless you like making old cars go really fast!

- **Renovation** is how the home looks.
- **Retrofit** is how it performs.

It doesn't have to be one or the other – we need both. Who doesn't love a beautiful, comfortable home that welcomes you at the end of a tough day? Who doesn't want plenty of storage space, good lighting and a good utility room? I know I do.

Then again – how marvellous to be oblivious of the weather outside, to have low energy bills and a home that's worth a load more money just because it's so cheap to run?

Renovation is something builders are very familiar with. In fact, this is the bread and butter for many companies, both small and large. Walk along any street and you'll see houses with skips outside as people work on making the best of the space they have. Building a kitchen extension, going up into the loft, adding a new bathroom…

It makes home comfortable, easier to live in and more attractive overall. A home 'with potential' is one that can be renovated and, as a result, they are much sought after.

Just look on Instagram or TikTok and you'll see endless reels of people going into their 'fixer-upper' for the first time. The focus is always on renovation – they rush in and:

- remove carpets
- knock down partition walls
- pull out kitchens
- rip up the garden
- open up fireplaces.

I'm more used to it now and manage not to throw my phone across the room. I scroll past questions about which Farrow & Ball paint colour they should choose and the fascination of sanding old floorboards. But all the time in my head is a list of what they should really be doing.

Problem is, retrofit is less sexy and not as instantly satisfying as a new carpet transforming a boring room into something spectacular. It's only the likes of me that get excited about hemp insulation (but you're very welcome to join me!). Sexy or not, retrofit is key to remaining comfortable in this fast-changing climate.

WE NEED TO FUTURE-PROOF

Get the bones of your home right and everything else can follow. Focus on getting the basics right:

- add insulation before putting plasterboard back up
- when floorboards are exposed, lift them and suspend wood fibre or hemp underneath

- when your old gas boiler needs replacing, switch over to an air source heat pump.

You need a home that can manage the climate changes that are already with us. You want to live comfortably without having to manage massive energy bills while creating a home that's warm in winter and cool in summer.

All of these challenges are addressed by retrofit. And you don't have to give up on your renovation. Final dressing of the house may take a little longer for budget reasons, but all the beautiful elements can be added whenever you're ready. Making your home airtight is a one-off that needs the building to be down to the bare bones, so that needs to take priority.

Why don't builders know this?

If you think about the way our world has functioned, it's not too difficult to understand why builders aren't up to speed with future-proofing.

I grew up with a coal fire in the main room of an otherwise pretty cold house. It was uncomfortable, but we lived accordingly. Most of life happened in that main room. We wore vests, liberty bodices, woolly jumpers, long johns... all the warm clothes we had. We used hot water bottles with plenty of blankets and an eiderdown on the bed at night. Baths were taken once a week because it was just too bl***y cold and the rest of the time a quick wash was sufficient – a lick and a promise, my mother called it.

Then came central heating. Strange to think that we've only had that since the early 1980s. As soon as we could warm the whole house with cheap gas energy, we got used to heating all the rooms, taking daily showers with the endless hot water and wearing T-shirts in winter.

The fact that we could manage with only perfunctory insulation was down to the gas boiler running at 65°C on thermostats, so we could just heat the house when we needed it.

Moisture was dissipated by the numerous draughts, which were integral to both old and new houses, so ventilation wasn't on the agenda. And if the natural air flow became too much, we just turned the boiler on for a bit longer.

So builders could keep doing what they've always done. As long as houses were warm in our damp, cold winters – and gas boilers guaranteed that – there was nothing to worry about. Certainly there was no need to add cooling for our unpredictable summers.

So all was well – as long as we ignored the rants of a few climate nuts.

That was me ranting!

I was one of those climate nuts but even I didn't realise the significance of our houses. Leaky homes are a major contribution to climate change – 20% of global emissions – which means they are inadequate for what we actually need. These days we are living in a world with:

- increased air pollution – inside and outside
- unpredictable weather events unlike anything we've known before
- sky-high energy bills that affect our ability to live comfortably
- condensation and mould that's a health hazard
- food shortages and price increases.

The way we build and renovate our homes must improve and it begins with mindset. Humans love the familiar, which is fine when life's predictable. But the pace of change is getting faster so we need to respond with equal speed.

For builders this means three big shifts:

- recognising that the houses built now are not good enough
- accepting the importance of sustainable heating, renewable energy and house 'bones' that can withstand challenging weather
- improving existing homes by embracing the importance of retrofit.

In particular, this means learning about the needs of old homes so they can become efficient and healthy eco homes for the 21st century.

It's estimated that 80% of the houses that will still be standing in 2050 have already been built. So the good news is that old solid structures are clearly standing the test of time. The bad news is that they're just not in the shape we need.

It's estimated that we have to retrofit one house every 90 seconds if we want to reach our 2050 climate targets!

Why are builders slow to get on board?

Don't want to be bothered? Don't know what they don't know? Don't see why it matters?

It's easy to see how it happens. It's so much easier to change when you believe it's worthwhile and you stand to gain. When someone comes along and tells you what to do, it's an invitation to resist. Then we're dealing with pride, resistance, complacency – all the vagaries of human emotion.

And we homeowners are part of the problem. We accept what the builders say. We bring them into our homes as an authority on this renovation business and want to keep them happy. As the customer we pay the bills and accept the builder as the expert/specialist. It's easy to get into that 'top-down', 'parent–child' situation where we lose our own authority and go along with what we're told is best.

But they are the experts

Indeed, builders are expert in what they do. It's what they don't do we need to be bothered about.

It's a rare builder who'll talk to the customer about:

Energy efficiency: imagine what a sell that would be! Imagine being told: "For a relatively small amount of additional cost now, we can reduce your energy bills for the rest of your time in this house, make you considerably more comfortable and increase the value of your home even more than the planned work will do."

The different materials we could use: what might work best for your house and the advantages. In particular, how certain insulation materials will keep you cool in summer as well as warm in winter. And the materials your old house actually needs, if you want improvements without laying in store more problems.

Alternative plasters: the different plasters that will help manage humidity, add insulation and be healthier for your home.

The ventilation needed: telling you that building regulations' ventilation measures are not enough for a healthy home.

Would that interest you? Would you be up for finding out more? Maybe extending the reach of the work you have planned?

My guess is you would. Sure, it might mean putting some of the style changes on the back burner for a short time, but as soon as you understood the long-term effects of what could be achieved, you'd be on board.

What's needed is a compelling reason for change and that sits with you and me. We hold the purse strings, so it's for us to educate ourselves about what we really need so we can:

- challenge the thinking of the standard builders in our area so they have to think more widely
- seek out specialist eco builders who really do understand the opportunities of home improvement
- employ a retrofit coordinator* who can tackle the builders for us.

Until we get our act together as customers, there's a real risk that any work on our homes will store up problems for the future.

We're trying to save energy on our own

It's not only the builders – we all need to learn about retrofit.

Since energy bills have hit the roof, people are taking matters into their own hands and finding ways to stop heat loss by closing off draughts so they can be warmer and pay out less.

Sounds sensible but we're forgetting – or we just never realised… Those irritating draughts serve a purpose – they manage the humidity in the home, so we don't get condensation and mould. As soon as we close them up, we risk moisture dripping down the walls.

So we all need to learn more about how houses work and what the new 'normal' requires of us.

WHERE TO START

When it comes to the 'how' of retrofit, you have two main options:

Whole house: you can grasp the nettle and do the whole place in one go. You might live in the mess or move out, whichever suits your family needs and budget.

We chose to live in. Partly to put all the money into the house without having to pay rent and partly to keep an eye on what was happening. But then it was just the two of us – no kids to factor into the equation – and we'd had builders in loads of times before so had a good idea what was in store.

One room at a time: you can work on one room at a time, completing the whole process and gradually building the thermal envelope (see below). This works well when you have to manage the family, balance the budget and balance your energy. It reduces the mess overall, but it does go on for longer.

We did a combination of the two. A straightforward extension morphed into a retrofit – not at all what had we set out to do – but since we were in a mess and down to the bones in places already, we decided to keep going and do it right. However, living in meant we needed a bolt hole, so we kept our bedroom and front sitting room for our own use. These were done later on when the dust had settled – literally – and we had the appetite again. Front room about one year later and the bedroom another year after that.

What is the thermal envelope?

Look at me – slipping in jargon without realising!

The thermal envelope is the barrier around your house separating inside from outside – the walls, floors, roof and windows all contribute to that barrier. There is always a thermal envelope, it's just not always very efficient.

Until recently most of us accepted the barrier as it was. It's been leaky and draughty because of inadequate windows, gaps around doors and floorboards, open chimneys, poor-quality building, but we managed. Energy was cheap, so who cared.

The thermal envelope we need now has to be much more efficient. We need the inside of our homes to maintain a comfortable climate regardless of the weather outside. In fact, that's an easy measure of your existing thermal envelope – how conscious are you of the weather? If you only know what it's like when you go outside, then you've got a good home for now and the future.

The vast majority of homes – regardless of when they were built – need to improve the envelope. That requires a building that is well insulated, with no air leakages and good ventilation. But more of that to follow.

MODERN BUILDING

Given that retrofit is about putting in what wasn't part of the original build, you could assume that it won't be needed for modern buildings. We'd hope that they were built to last with good standards of energy efficiency. Sadly this isn't always true.

Any home with an EPC (Energy Performance Certificate) of C or below is considered in need of a retrofit. New homes should be rated B or above – with the emphasis on 'should'. Houses built after 2012 should be energy efficient – if it is a quality build – but anything built before then will need some form of upgrading.

How come so many of our newer homes are in need of care? There are a number of reasons:

- **Cost:** when the call is put out for bids on new builds it's generally the cheapest that wins. What's not taken into account is the long-term expense of choosing cheap; lower-quality materials and less-skilled labour result in lower standards.

- **Built in a rush:** part of securing the winning bid will be to deliver quickly. This will lead to shortcuts and diminished quality control.

- **Quality of materials:** modern building materials are not nearly as durable, long lasting or sustainable as natural materials. But they are cheap and easy to access so fit perfectly with modern requirements of fast delivery.

- **Expectations:** people move home more than in the past. If they're not staying there's less concern for quality. They just want to know the house will sell well so they can make some money and buy bigger. Building for the long term is no longer valued.

- **Lack of skills:** not enough people have been trained to do a skilled job and those who are skilled are ageing so leaving the workforce.

All these factors together mean that the age of a property is no longer an indicator of what needs to be done. I recall seeing a thermal camera video of new homes that showed heat streaming from gaps in the eaves and around all windows and doors. It must be immensely disappointing for new homeowners when they go into their first winter with huge bills to pay just to stay warm.

WHAT ABOUT OLD HOMES?

In contrast, no one's surprised that older houses need improvement. There are loads of builders out there who've made a good business out of renovating old homes. Unfortunately, there are not enough builders who recognise that old properties have special needs, so they leave the home looking good for the short term while all sorts of problems are just biding their time.

They need to recognise that 21st-century materials won't work for a house built before 1940. But this information hasn't found its way into the psyche of the building profession. They continue with what they know, regardless of the needs of the building. At times this will even mean pulling out whole areas of the original house to remake it in the form of a new build – like filling the standard Victorian sub-floor void with concrete.

How old is my home?

If your house is a solid brick construction, has a floor suspended over a gap of varying depth, (sub-floor void), old-fashioned air bricks around the base of the building and a fancy pattern in the bricks rather than the straightforward stretcher bond*, then it will have been built before 1940.

Knowing this is very helpful. Houses built before 1940 were built with vapour-permeable materials, so any work done to the original fabric must also use vapour-permeable materials. It's the only way to avoid problems caused by condensation, mould and damp. There's an exciting array of options that also provide better protection against heat in the summer (see chapter 11), but it's more likely that your builder will default to their beloved PIR – a petrochemical-based insulation material that is not vapour-permeable (See chapter 6).

All those distinguishing features of the old house – fireplaces and chimneys, ventilation grills and sub-floor voids under the floorboards – were there for a purpose. They did a great job of letting in fresh air and managing moisture.

That combination of vapour-permeable materials and plenty of fresh air meant people in days gone by had few problems with condensation. But then we came along, closing up the gaps, using modern materials and replacing old windows with plastic ones, cutting out the draughts and inadvertently creating more problems than we solved.

IT'S ALL DOABLE

None of that needs to be a problem. All it needs is someone who understands the implications of building styles, who has knowledge of natural materials and the ability to explain and talk about what's needed.

It may take a bit of work to find the right people and it definitely means getting your head around some of the basics. But you only need enough knowledge to ask the right questions and make sure you get what you want.

A standard to help: PAS 2035

There are moves, in the UK anyway, to set a standard for retrofit. This is PAS 2035 *– Publicly Available Specification 2035 – a recognisable quality standard for the retrofit and energy-efficiency sector of housing. (See chapter 3.)

Unfortunately this is only a requirement for projects being delivered with government funding. So if you live in a home that fits the Warm Homes: Social Housing Fund*, Energy Company Obligation or Local Authority Delivery Scheme you get some cover from PAS 2035 because the building companies will have to comply.

For the rest of us, we can refer builders to PAS 2035 but there is no obligation for them to work to the standard. That being said, we can certainly employ a retrofit coordinator - a role designated by PAS 2035 to help maintain the standard. Find the right person to suit you and they will:

- survey your home from an energy-efficiency perspective, using information about your energy usage, state of the building and energy requirements
- provide a retrofit plan that can be executed in one go or split up into stages
- work with you to identify appropriate builders for your project
- project manage the work for you
- manage any possible risk factors with the builder.

This is a great compromise. If you don't trust that the available builders understand enough and you're not confident you can set the standard you want yourself, then a retrofit coordinator might be just what you need.

WHAT RETROFIT INCLUDES

Your aim is to:

- keep your house at a steady temperature throughout the year
- manage the moisture produced by living and weather through effective ventilation
- heat and create energy to do both of the above.

The three elements work perfectly together to provide a home that's comfy, cheap to run and ready for an uncertain future, while being healthy and safe for your family.

Keeping your home at a steady temperature

Running at a steady temperature is all about insulation and airtightness.

The one we talk about most is insulation. When grants are given by government it's generally for insulation, so we've heard plenty and think we know what we're talking about. But I can tell you there's a whole world of insulation materials out there that most people know nothing about – and they're considerably more interesting than the standard PIR most builders use.

Insulation is like putting on a warm jumper to go for a walk. It wraps your home up in a thick blanket, providing a worthy barrier between inside and outside.

When talking retrofit and healthy home, the type of insulation used is important. Sadly the synthetic insulations favoured by the majority of builders are energy-greedy. They also give off (off-gas) volatile organic compounds (VOCs). This is a problem for three reasons:

- VOCs can cause respiratory and skin problems.
- In rigid forms of insulation, off-gassing can cause shrinkage, leaving gaps for air to come through.
- Synthetic insulation is not vapour permeable, so its use can lead to problems with condensation.

These problems are avoided by the use of natural materials: wood fibre, sheep wool, recycled denim, hemp, sisal, cork – there is a fascinating list of options, but it's a rare builder who will suggest these to you. Which is a shame, because they have a number of advantages:

- they are naturally occurring, so don't involve ingredients that require mining, high energy usage or oil-based components
- they are sustainable, so reduce the impact on the earth – e.g. sheep's wool is grown and shed each year, hemp grows two crops a year and wood fibre is made from the scraps left over when everything else has been used
- they are vapour permeable so they manage moisture well, reducing condensation.

Airtightness is the companion to insulation because one without the other will leave you disappointed.

However well insulation is installed, it's really hard to avoid gaps. And air is really pesky – it will get in anywhere. So the combination of an airtightness membrane installed alongside insulation gives you the best of both worlds.

In retrofit your aim is to close off uncontrolled air. The airtightness membrane is like putting a windproof jacket on over your woolly jumper before going out in winter. It stops the cold air passing through your clothes, so you retain your natural heat.

Numerous homes in the UK are plagued with draughts that cool down the heated air. If you have a draughty home you'll probably know where they come in, but if you need to check, a thermal imaging camera* is a good way to do it.

Windows are part of the plan for keeping your home at a steady temperature throughout the year. Double glazing is often one of the first upgrades people make. We know quite a lot about this because of the constant advertising for double glazing. (see chapter 14)

Managing moisture

Moisture creates more problems in our homes than anything else, yet ventilation is rarely included in discussions about making a home energy efficient.

By insulating and making airtight, you've wrapped your house up to be nice and warm, but you've also taken out the uncontrolled air – in other words, you have created a sweaty box.

Let me explain: a family of four produces 20 litres of moisture per day. Just by living. Cooking, showering, sweating, breathing... And if that moisture has nowhere to go it'll condense and drip down the walls.

Even worse – a wet wall is a cold wall. And as your carefully warmed air hits the cold surface then there is even more condensation. It's a terrible cycle that leads to mould and all manner of illnesses.

Not to mention that a wet wall is much harder to heat.

But all is not lost – you just need to balance staying warm with managing the moisture.

There are a number of ways to do this. Everything from opening the windows up to full-blown mechanical ventilation heat recovery (MVHR). There is plenty more detail about this in chapter 10.

Breathability is the other way to account for moisture and is most relevant for old properties. Just like ventilation, you ignore breathability at your peril. And as with ventilation, the risk factor is condensation – either inside your home or inside the walls of your home.

Breathability is the passive movement of water vapour through the actual structure of your building. The degree to which your home is breathable depends entirely on the materials used. Vapour-permeable materials enable the free movement of vapour from inside to outside and vice versa. (See chapter 10)

HEATING AND ENERGY

Heat pumps are the main form of sustainable heating for the home. When your gas boiler dies, that's your opportunity to move to a heat pump. An essential part of future-proofing – we're all expected to switch over by 2035 – they are more efficient and cheaper to run.

Heat pumps are subject to prejudice about new technology. They're like Marmite – people love them or hate them. Which makes for all manner of stories about how great they are and how dreadful they are. You'll probably be told that a heat pump:

- won't keep your home warm
- will cost a fortune to run
- will be noisy
- will annoy the neighbours
- won't give you enough hot water.

My advice: if you want to know more about heat pumps speak to someone who's got one. Ignore the press. For all sorts of reasons they prefer to be negative on this one. In reality when things do go wrong there are good reasons why and none are to do with the heat pump itself. Most often it is to do with the installation or our mindset. (See chapter 12 for more.)

Generating energy is an exciting part of this retrofit process. It's great to be self-reliant, fuelling your life in a renewable way.

Solar and wind are the main options. I'd love a wind turbine and I did ask about it once, but we'd need a huge garden to accommodate it – the height of the turbine plus 10% distance from the house. Plus it's costly for the amount of energy produced. So I'll leave that one to the energy providers.

Solar, on the other hand, is easy and relatively cheap to install – and getting cheaper by the day. All you need is a roof and you're on your way. Ideally the roof faces south, but east/west also work well. Only north is questionable and probably not worth the investment.

More detail about all of this in chapter 13.

Summary

- Retrofit improves the performance of your home, making it energy efficient. Builders won't automatically include it so you have to say what you want.
- You can retrofit the whole house in one go or take one room at a time as your budget allows.
- PAS 2035 is a standard that helps you identify what to do. It is used in social housing but is a good guide for any homeowner.
- A retrofit coordinator can work out a plan for you. Be sure to say that you want natural materials before they start. They will all think energy efficiency. They may not all think about eco materials.
- The aim is to keep your home at a steady temperature all year round.

To find the information indicated by the * follow the QR code.

NOTES

CHAPTER 3
RETROFIT BENEFITS THE BUILDER DOESN'T KNOW ABOUT

Let's cut straight to the chase and look at data from my own experience.

NB: Because the price of energy is so volatile, we used kilowatt hours (kWh) for all our measurements. This gives us a real comparison without having to take price rises into account.

Energy use

In the last full year before we started the work, we used 25,500kWh to power the house and the car.

In the first full year after all the work was completed, we used just under 6,000kWh.

In subsequent years we have recorded between 6,000kWh and 7,300kWh per annum.

That's a 75% reduction in energy usage.

Energy Performance Certificate (EPC)

An EPC shows the energy efficiency of a building. It gives a rating from A (good) to G (poor). Sadly it's still rare for a building to receive an EPC of A. D and below are all considered in need of renovation or retrofit.

Our Victorian terrace originally had an EPC of D. Once all the work had been completed, we were awarded an EPC of B. This is very unusual in a Victorian house, so we were really pleased.

House value

This is a favourite question: 'Is the house worth more as a result of the work you did?'

I had two valuations done. One of the agents understood nothing about retrofit so assessed it purely on rooms, location, amenities, internet points etc. The second agent understood what we'd done. He valued the house at 'at least' £90K more than the equivalent house without a retrofit.

Since then he's told me the difference would be even more now, because so many buyers are looking for energy-efficient homes.

What was the cost?

We worked out the additional cost on top of the work that was being done already, so:

- additional labour to insulate and make airtight in the rest of the house
- cost of materials – insulation, airtight membrane and tape etc (excluding what was already included for the extension)
- single-room heat recovery units – three at the outset

(we've added more since, but they are not included here)
- air source heat pump (amount paid after the £7,500 government grant)
- eight solar panels on the house.

For all of the above we paid an extra £25,000.

So £25K to increase the value of the house by £90K, reduce our energy bills by 75% and leave us feeling warm and comfortable.

It amazes me that builders don't see the sales potential in this process. Imagine if a builder came to you with a quote for your home improvement with an addition that said:

"For £___ extra we can make your house energy efficient, cut your electricity bills significantly and make your home worth a lot more money."

Would you go for it? I certainly would.

When considered in the context of a new kitchen extension an extra £25K is relatively small. But it'll deliver a major impact. Now that's what I call a unique selling proposition. We'd have bitten the builder's hand off if he'd known to suggest it.

But it's highly unlikely you'll be offered this as an option, so it's going to be up to you to ask and insist.

We could have done better

Despite the benefits we've gained, I now know that we could have done better. Which is even more amazing – we could be saving even more than 75% of our energy usage!

One way to do that is by going for an EnerPHit-level retrofit ('Energy Retrofit with Passive House Components'), which is the gold standard of retrofitting.

RETROFIT BENEFITS THE BUILDER DOESN'T KNOW ABOUT

What's a Passive House?

Passive House* is a building standard that puts energy efficiency first in order to minimise the heating and cooling needed. It's achieved through a combination of design and construction techniques.

I was amazed the first time I visited a Passive House. Whatever the weather is like outside, you are comfortable. Now that may not sound so amazing until you realise that the owners pay no energy bills at all. In fact, they make substantial money from selling solar energy to the grid.

The Passive House is totally airtight, thoroughly insulated and well ventilated, with triple-glazed windows, so it holds to steady state whatever is going on outside. No heating is needed, other than that created by the occupants. The occasional use of a hairdryer, cooking dinner, boiling the kettle, having a shower – all give off heat that the building can hold onto because of the level of insulation and the structure of the system.

Clearly this requires great precision in the building process so calls for sustainability experts, who are few and far between just now. It can also only be achieved in a new build where the builders have total control over the process.

Which is where EnerPHit comes in

But that doesn't mean those of us in existing houses can't aspire to high standards of energy efficiency. EnerPHit* is a standard that's as close to Passive House as is possible in an existing build.

By the time John and I learned about EnerPHit, it was too late. You have to begin the process in the planning phase and we were way beyond that. But if you are keen to get the very best out of the retrofit, then explore EnerPHit* and consider it

AECB CarbonLite

CarbonLite* is a standard developed by the AECB (Association for Environment Conscious Building) to help homeowners and builders create energy-efficient homes without the extra expense of the EnerPHit system. It includes taking a retrofit one step at a time with a different standard for each step, so acknowledges the challenges with cost and the needs of the regular homeowner.

PAS 2035

PAS 2035* is the British standard for energy retrofitting of homes. It provides a whole-house framework, including health, moisture risk and ventilation, as well as insulation. It sets out the complete process, from assessment to monitoring. It is also why we now have retrofit coordinators to turn to (see chapter 8).

Eco house

So whether you follow a particular standard, work with eco-inspired professionals or educate yourself, you can create yourself an eco house: a home that is energy efficient, sustainable and reduces your impact on the environment. To do this means exploring the world of 'healthy home', a phrase that is becoming increasingly important in our modern world, but is not yet on the radar of most builders (see chapter 16).

Gradually people are coming to realise that the homes we live in, the products we use, the way we live, even the clothes we wear have a direct effect on our health and the health of the planet. We just need to get the message across to builders and building professionals.

RETROFIT BENEFITS THE BUILDER DOESN'T KNOW ABOUT

WHEN TO RETROFIT

There is a step before you even think about retrofit. Assess your house to see where the easy wins might be:

- A damp/wet wall is a cold wall. Check your guttering and downpipes for blockages that will cause an overflow.
- Make sure your drains are in good shape. Do any repairs ahead of any major work, unless drains are to be included.
- Look for gaps in the render or missing mortar – this may be letting in water, making the home cold.

Making these checks can reduce the risk of damp and cold. It can also highlight other jobs that need to be done by builders, which can be added into an estimate.

Take the opportunity

I see so many people posting on social media about their home renovation. They go back to the brick because the plaster is dodgy, then put up plaster board and replaster. It breaks my heart and I want to shout: Stop! Think! Insulate!

Going back to the brick is the messiest part of renovation, so make the most of it. Once you're at this point, it's minimal time and cost to make the exterior walls airtight and attach natural insulation. And the payoff over time is enormous.

Avoid renovator's regret

Building an extension, putting in a new kitchen or bathroom, changing the floor… Every improvement can be adapted to include retrofit elements while you're at it.

Miss it out and you could be sitting in a beautiful home with cold feet and a huge energy bill. There is nothing worse than living through a build, enduring the mess, paying out the money and then realising you missed a trick. And a trick that would have made a massive difference to your life going forward.

Summary

- Retrofit will save you loads of money on energy bills, you'll be more comfortable and your home will be worth more. And you will help the planet.
- If you are extending or improving your home this is the perfect time to add in retrofit. The extra cost is small in comparison to the work you'll be doing.
- You can add design features later; retrofit takes place alongside the messy stuff so take the opportunity while you can.
- There are measures to help you that outline what is needed. You can use these if you want to or just take the information to share with your architect, designer or builder.
- Don't experience renovator's regret – it is really upsetting to have a beautiful house and cold feet.

To find the information indicated by the * follow the QR code.

NOTES

CHAPTER 4
HOW CONSTRUCTION IMPACTS THE CLIMATE

I've been mithering about climate and the impact of human beings since the early 1970s, so it's easy to assume I know all about it and what the problem is.

Writing this book has made me wrap my head around exactly why construction has such a poor track record when it comes to carbon emissions and damage to the environment. And that's been useful in taking me back to absolute basics and defining the issue – just to make sure we all get it.

This chapter, then, explains what the builder doesn't understand about the climate impact of construction – how construction is harming the environment and how we can change that.

THE BOTTOM LINE: WE ACTUALLY NEED TO STOP DIGGING

Let's start at the beginning. Carbon is a fundamental component of all life and is present in everything that lives or has ever lived. When anything containing carbon is burned, that carbon is released and it combines with oxygen to create the carbon dioxide molecule (CO_2). Carbon dioxide is also produced by decaying matter and by our breathing.

So far so good

But carbon dioxide is a greenhouse gas. This means it absorbs heat from the sun's infrared radiation bouncing back off the surface of the planet and traps it in the atmosphere. The concentration of CO_2 in the atmosphere has a direct bearing on the average temperature of the planet.

Over millions of years, CO_2 levels have tended to follow other naturally occurring planetary warming or cooling influences, contributing to ice ages, interglacial periods and greenhouse periods.

Today, though, thanks to human-induced emissions, CO_2 is very definitely leading, not following, and our modern way of life is causing atmospheric warming at a far faster pace than at any other time in earth's history.

So in short:

- carbon stored safely in the earth has no impact on our atmosphere or our climate
- but carbon dug up and burned releases carbon dioxide (CO_2)
- CO_2 absorbs heat and traps it in the atmosphere and that causes the temperature to rise.

NOW WE NEED A BIT OF HISTORY

Coal is a naturally occurring material, made up of between 60% and 97% carbon. It was used for thousands of years for heating and industrial processes, but in relatively small amounts because of the difficulty of mining deep seams. But with the Industrial Revolution and the invention of the steam engine, its use increased exponentially. It was easy to use coal because:

- coal was available for use as soon as it was dug up, whereas wood has to dry out before use
- coal burns more consistently than wood
- coal was cheaper to transport – a wagon load of coal contained more potential heat than a wagon load of wood.

Coal was formed from the breakdown of plant and animal matter. It was nature's way of locking up all that carbon so the atmosphere wasn't affected. You see, when left to its own devices, the earth has a way of managing and holding life in the ideal balance.

With our big brains and bright ideas we think we can just have what we want and all will be well. Sadly with each new idea we upset that delicate balance between life and earth a little bit more.

It was a good idea at the time

The use of coal and production of steam was such a good idea. And it could have stayed a good idea if we'd been a bit more circumspect.

Steam was just the beginning. The internal combustion engine, developing industry and major growth in population have required us to burn more and more coal and oil, producing more and more CO_2. All of which has resulted in an increase in average surface temperature of more than 1.3° C. We are already on track for it to increase by at least another 1.4° C by the end of the century.

A bit of perspective

1.3°C doesn't sound bad. After all, sitting in the garden having a cuppa with a few extra degrees might be quite nice.

Sadly what we're talking about is average temperature. But even then, it's not that much. Is it?

To get some perspective, the average temperature during the last ice age was only 4°C cooler than we are today. Four degrees C doesn't sound that much, either, does it?

But if just 4° cooler created an ice age, what will 3° of heating do? Earth has never been more than 2° above pre-industrial times in the past three million years.

Now it starts to feel very serious. It means sea level rises, extreme weather events, crop failures – the list goes on.

WHY IS THIS RELEVANT TO CONSTRUCTION?

The building and construction sector has made a major contribution to our problems with climate. Estimates range from 37% to 50% of greenhouse gas emissions, so it is certainly a problematic area. Efforts are being made, but it's hard to know whether real change is happening or if it's just ticking boxes. Noise pollution, air pollution, water pollution, soil compaction and erosion, destruction of natural habitats…

We need to consider:

- the materials we use and their impact
- waste production as a result of our build
- toxins in the beautifying of the home.

Materials we use and their impact

Out of habit, the building trade uses materials that are detrimental*. It's what today's builders are used to:

High-carbon materials:

- **Cement and concrete:** 5–8% of greenhouse gas emissions because production is energy-intensive using high-temperature heating
- **Steel and aluminium:** large amounts of energy used in extraction, processing and manufacture. Not to mention the CO_2 from coking coal used to produce the steel in the first place.
- **New bricks:** firing bricks at 1,000°C consumes vast amounts of energy and releases carbon.

Toxic and non-renewable materials:

- **Polyvinyl chloride (PVC):** used in pipes, flooring and window frames, PVC contains harmful chlorine and petroleum-based compounds, all releasing greenhouse gases.
- **Synthetic insulation:** expanded and extruded polystyrene (EPS/XPS), polyurethane (PUR/PIR) and phenolic foam are oil-based, and their production releases significant greenhouse gases.
- **Fibreglass:** fibreglass manufacturing is energy-intensive and can contribute to indoor air pollution. The tiny fibres can also cause skin and respiratory irritation during installation.
- **Composite wood products:** plywood, particleboard and other composite woods often use formaldehyde as a binding agent that off-gases, releasing harmful volatile organic compounds (VOCs).

Wood itself is also an issue. Deforestation is a very real problem for the climate. Buying products made from new wood adds to that load – for the natural habitat, transporting the wood to the place of manufacture and any chemical treatment that is added to the wood.

Waste production as a result of the build

You only need to see the skips sitting outside houses as you walk around your neighbourhood to understand how much waste is produced. Discussions about home projects often include competition over who used the most skips.

Builders won't, as a rule, take the time to sort materials into what can be recycled and what absolutely must go to landfill.

One carrier bag of waste

I met the owner of a company that did replacement windows – a business that traditionally has high levels of waste thanks to all the old windows removed. He was keen that the company should take care of the environment so they sorted out everything that could be recycled and did their very best to reuse materials wherever they could.

End result? One standard carrier bag of waste per week from the whole business – including the standard office waste and lunchtime rubbish.

We didn't do so well in our retrofit, but John was out every evening going through the skip to take out what could be recycled, given away or reused. I met some fabulous people when passing on our waste to someone who really wanted it*.

Toxins in beautifying the home

As the work on the house draws to a close it's easy to take your eye off the ball. You just want to settle down and be comfortable again. Sadly there are still issues to address:

- **toxic paints** that reduce the vapour permeability of the walls, and trap condensation between the paint and the wall, as well as off-gassing. Not to mention the damage done in manufacture
- **furniture:** VOCs are plentiful in furniture, mostly in the form of glues and varnishes
- **vinyl and laminate flooring:** often petroleum-based, using significant energy and releasing harmful substances during production, and off-gassing at home.

CAN HOMEOWNERS HELP WITH ANY OF THIS?

As homeowners we find ourselves in a strange position. We need to adapt our homes to manage a much wider range of temperatures and climate conditions. However the way we set about this could become part of the problem.

For example, if we future-proof our homes using synthetic materials that require mining, drilling for oil and the use of chemicals, then we contribute to the process that requires us to future-proof in the first place – and the wheel goes around one more turn.

The answer is to take care with the materials we use in renovations and retrofit. There are plenty of options out there for you to choose from.

Don't expect the builder to tell you anything about this – unless you find specialist eco builders. You'll need to do the research yourself. While being highly efficient and effective materials, these are not yet on the radar of the majority of builders or building companies.

So look for:

Low-carbon/bio-based materials: consider timber, which sequesters carbon, or recycled concrete. Other bio-based materials like hempcrete, mycelium bricks or bamboo can replace certain structural elements.

Natural insulation: choose natural alternatives like sheep's wool, wood fibre boards, cellulose, hemp – see chapter 11 for more. These materials are renewable, biodegradable and offer comparable or even improved performance compared with synthetic options.

Recycled and reclaimed materials: use reclaimed and sustainable materials wherever you can: reclaimed steel, glass, wood or brick to reduce the environmental impact of extraction and manufacturing. Sustainable timber will hold carbon for very long periods of time, or use recycled wood, which also stops it releasing its carbon through burning or decay. (see chapter 15)

Low-VOC paints and finishes: select paints, adhesives and sealants with low or no VOCs to improve indoor air quality. (See chapter 11)

Innovative concrete alternatives: for specific applications, new materials like ferrock (made from steel dust), ashcrete (using fly ash), limecrete, blown glass and geopolymer concrete offer lower-carbon solutions, or you could use screw piles instead.*

It takes some work – a lot of the answers are here in this book – but it's a fascinating journey. Not to mention that you are working for the generations to come – your future grandkids will thank you.

Junckers gym floor

Once you start to look you'll find all manner of interesting ways to complete a build.

We built a garden studio with a south-facing roof for solar panels. It was all done without brick or concrete and we were determined to keep it sustainable all the way through. We wanted a wood floor and John managed to find a company selling second-hand gym floors.

Seventy-year-old solid beech floor. I've never seen anything like that in a flooring shop or website. This was something special. You can even have some of the old gym markings left on if you choose to*.

It was a challenge to lay down. If we'd paid a bit more money, it would have been much easier, but John was determined not to give up on it. He made his own tools and worked out a system for how to get it down. It's a work of art. At least, it will be when the final bits of skirting board go down!

There are trades out there who will do it for you – so it doesn't need to be a lot of work. And you'll never find a floor as good anywhere for the money.

It will see us out!

THIS ISN'T ABOUT THE PLANET

It's easy to lay all of this at the feet of the planet. Let's be brutally honest here. The planet is going to be fine. All this has happened before and she's survived.

History tells us that it's life that will suffer. Animals, plants and humans can't manage the extreme temperatures that we're creating. We just won't be able to sustain life as climate events become increasingly extreme. It may not be forever – species could return in time, as conditions change again, but we're talking millennia.

So this is about all life, including you, me and the grandkids to come. Deciding to manage the CO_2 in the atmosphere is not altruistic. It's a selfish, self-focused decision. It's us looking after our own needs and the needs of future generations, as well as other forms of life.

What is the way out?

We know what to do. Everything we need is already out there. We could begin now to live a more balanced life, in harmony with nature.

But we have to stop digging:

- Stop using coal to create energy – use wind, water and solar instead.
- Reduce our consumption, so reducing landfill – focus on quality over quantity.
- Stop driving internal combustion engine cars – switch to electric vehicles, walk, cycle, use the train.
- Eat fewer animal products – eat plant-based and leave more land to return to nature, sequestering carbon.

- Become more energy efficient in life overall and at home.

What about climate-friendly construction?

There is plenty of good news on this front and – you know what I'm going to say - we homeowners are key!

There are loads of building materials out there that are significantly more environmentally friendly.

And not only eco-friendly – they are also better for us. They don't contain any of the toxins found in synthetic materials, they do their job better and they can have a positive effect on the climate.

Hemp is the perfect example. It grows so quickly that it can be harvested twice a year and it holds a load of carbon. By using hemp insulation you can make your home a carbon sink – locking in the carbon so it doesn't add heat to the atmosphere.

There are numerous other natural insulation materials that also hold onto carbon so you can future-proof your home without being part of the reason why future-proofing is needed. You also need them if you want your house to stay cool in summer – something we are only now needing to think about. Synthetics will never do that. (See chapter 11.)

By installing a heat pump, you can heat your home from the air, rather than using oil or gas and releasing their long-held carbon into the air. (See chapter 12) And if you can fit solar it can work in tandem with your heat pump, allowing you to live off the sun – no carbon involved at all. (See chapter 13.)

Each day new materials come along that use recycled materials or exciting new products – like leather-type material grown from mycelium, slippers made from vac dust and 3D-printed sand making home furniture.

All we have to do is stay open to change, try new ideas and set the path. The sooner we can make this way of living mainstream, the sooner we can begin to heal some of the damage we've done.

It's getting very late. But we can still do something about it. And starting with your home – where you need make no sacrifice, but just get warm, comfy and better off – is the perfect first step.

We can do this – we just have to stop digging.

Summary

- Every time we dig something out of the earth, we release carbon, which forms carbon dioxide (CO_2). CO_2 causes more heat to be held in the earth's atmosphere for centuries.
- As the process progresses, it gets worse. As the earth heats, the ice caps melt and permafrost thaws, releasing even more methane, which causes the earth to heat…
- Standard building materials used every day in all forms of construction are destructive for the earth, but there are loads of options out there that also do a better job for us.
- The earth will survive, but life in all its forms won't – and that includes human beings.
- But there is a lot of hope – if we take action now. And one place for you to start is with your home. Cut your energy costs, store carbon with your choice of materials, use the air and the sun.

To find the information indicated by the * follow the QR code

NOTES

CHAPTER 5
HOW TO MANAGE THE BUILDER

Deciding on a builder is a big decision and not to be taken lightly. In fact, alongside the cost of home improvement, this is one of the main reasons people hesitate. It's just so difficult to know who to trust.

'You won't believe what my builders did' is the stuff of dinner party entertainment. There will always be those with lurid horror stories that make you want to put your head in the sand until it's over. Others will be overly optimistic, with nothing bad to say about any of it.

Don't expect the builder to be much help, either. This is the stuff of routine to them, so they assume everyone understands. They also see things from their own perspective, so it may not occur to them that you need some advice or information. They'll focus on the details – plans, timing and cost. This might increase your confidence in the short term, but understanding more about the different elements in play will be much more helpful.

What you actually need is some direction about how to sort out the day-to-day with your builders without getting into hot water.

So here goes.

Two worlds collide

When you plan changes to your home you'll think about what you want it to look like. The reality of dust, mess and noise is daunting, so you'll imagine yourself in your new kitchen, sitting by the fire or showing friends proudly around your new home. You might worry about the cost and how long it's going to take, but overall you'll be looking forward to this new phase of your life.

The builder, however, will get straight down to detail. Can they manage this project? How will it fit into the diary alongside other work? Will it add to the bottom line? They'll be considering what trades will be involved, whether the planning permission is in place, how long it's likely to take. They might give a thought to whether they can get on with you, but that's more likely to come later.

Of course, neither of you will think to say any of that – you'll both just assume the other is on the same page, focusing on similar things. In fact, you could easily be talking at cross-purposes.

It's the same in every conversation in every situation. We assume we understand what's being said. We also believe we're making complete sense. Sometimes we're right, but sometimes nothing could be further from the truth.

When we enter into a romantic relationship, get to know a new boss or just make new friends, we accept that it takes time to understand the other person. We ask questions, listen to stories, find out what makes them happy or satisfied with our work. We want to get it right so we put the effort in.

It's not always the same with the builder. As customers, we expect them to read our minds. We hand over plans, describe what we want and expect that to be enough. As time goes by, we begin to realise that making assumptions was a mistake.

HOW TO CHOOSE YOUR BUILDER

For any project that's more than a few days' work, you're going to be spending a lot of time with this person, so you need to choose carefully.

The first challenge is finding builders to ask for quotes, so:

- Talk to friends who've had the builder in. Were they happy with the quality of the work; did they get what they wanted; did they feel OK with that person in their home? Ask them about the finish of the work and whether they would have them back.

- Look on local social media pages. Facebook often has a local renovations page where people will talk about the builders they've worked with. Focus on the pages that are just for customers – once builders themselves are included, the information becomes less reliable. Ask for suggestions and check out the names you already have; get other views on the people you're going to meet.

- Look the builders up in trade federations – e.g. Federation of Master Builders* (FMB) or National Federation of Builders* (NFB). Both groups vet their members and have a code of conduct.

- Take a look on TrustMark*. This is a government-endorsed quality scheme. Not every builder will be on there, including some very good ones, but it will give you a start:

Before you meet the builder, give them the basic information:

- the plans you have for the project
- the time frame you have in mind and any specific deadlines
- any challenges that you already know about – e.g. issues with the drains, problems with exterior cladding…
- what you want from the meeting – do you want an opinion; a rough guideline on cost; an indicator of obstacles…
- let them know if you want to speak with previous customers to find out about their work.

You'll save time and energy if you're honest about the non-negotiables, so always have a very clear agenda or brief to give to them. For example, if you have a specific deadline it's much better to say that rather than waste time meeting with someone who is never going to be able to achieve that date.

First meeting

First port of call, trust your gut. If you have an instant reaction against someone, see if it fades during your conversation, but never discount it. We can't get on with everyone.

Invite the person in, put the kettle on, have the usual opening chat. You're considering whether to let them into your home day after day. It's worth investing a bit of time.

When you're ready, talk through the information you sent. Have a notebook with you and write down questions as you think of them. You'll have a lot of thoughts going through your head and it'll be easy to forget. Find out what they think about the plans. Take them to the place in the house that's under discussion. Show them the areas that connect to the plans you've provided.

Let them know what materials you want to use and ask if they have done similar projects, using these materials, before. You need to know if this is the type of project they're familiar with.

Listen to their thoughts, suggestions and concerns. In my experience, most builders will have ideas for how it can be done differently and why some ideas won't work in practice. They may be right or they may just not like the idea. You have to work out which it is and place that alongside the reason you wanted to go that way in the first place. Then you have a choice to make.

This discussion will provide clues to what the person will be like to work with. Pay attention to the content and the style. Do you feel positive about the suggestions or is it putting your back up? Are they really listening to you and taking in what you want or just trotting out the standard information?

Most disputes are due to poor communication so do find out:

- who will be on site each day
- who is the best person to speak to when you have a question
- when and how often they will meet you to discuss progress
- how they prefer to handle any disagreements.

If the answers don't work for you, say what you want and see if they are willing to get on board. If so plan progress meetings to suit your needs and timetable so you can stay in touch with what's happening. Don't let the builder decide, otherwise they could just scoot off at the end of the day leaving you none the wiser on progress.

I'm just too polite

We've had enough work done on our house over the years for me to know that I get caught up in being polite. A lot of times this is fine – I don't always have to speak my piece. Sometimes it's fine to just listen to what someone else has to say and accept it.

Except when it comes to getting what we need. There are definitely times when it's vital to speak out and be heard.

I finally understood this when we decided to make our retrofit eco-friendly. Of course this took the builder out of his comfort zone – we were saying a hard 'No' to PIR, the insulation he used on automatic pilot.

He is a really good guy – very willing to talk through our ideas, speak out when he thinks something won't work and go with us when he is satisfied. But even so, there was plenty of teasing about the wood fibre, airtight membrane – all the materials we wanted to use.

Eventually it broke even my capacity to be polite and I got irritated. I told him: "Stop it. This is the way the world is going so you need to get on board." Then I made him a cuppa and we all got back to work.

Of course, builders are only human, so they'll be on their best behaviour when they first speak to you. You are a potential customer and they'll want to provide what you're asking for. At this point, the more you can demonstrate that you're an engaged customer who knows something about the process, the more seriously they'll take you. Of course they may decide they don't want a customer 'interfering', in which case you're better off not working with them.

This isn't easy stuff. I recommend that you speak to a few builders before making your choice. Each time you go through the conversation you'll be clearer about what you want, you'll recognise standard sales speak and you'll be more able to spot the person you can connect with. Trust your gut instinct – this person will be in your home for several months, so you need to be able to get on.

Tough conversations

It always happens. You'd be extremely lucky to get to the end of a building job without at least one delicate or difficult conversation. It will often be about money:

- they ask for more than you expected or extra jobs have been added without you understanding the extra cost (see chapter 9*)
- you don't think something has been done well enough and you want them to do it again
- you forgot to get money out of the bank on the right day.

No one likes having these discussions. And it's important to pick your battles. If something isn't that significant and you are happy to live with it, then let it go. After all, no one sets out to do a bad job.

On the other hand, some builders content themselves with a 'good enough' job and cutting corners can happen if they're in a hurry. Both of which need to be addressed. As a rule of thumb, if the issue is going to stick in your mind or irritate you every time you think about it or see the builder, then it needs to be discussed. If the work is ongoing, then clearing the air will be the only way to manage the rest of the process.

Resentment leaks and builds up with each minor disagreement. So you may need to take the plunge and sort out the issue for the sake of the wider project.

LISTENING IS KEY

Quality listening is the key to understanding and, above all, you need the builder to understand what you want. Problem is, they're used to giving advice. And, of course, you want them to be expert and know exactly what they're doing, but it does mean they won't necessarily pay much attention to you and how you feel.

So getting your needs met is down to you. List all your questions before you meet and work through them systematically. Remember that you're the customer and the point of every conversation is to make sure you get the home you want in a way that is safe and manageable.

Ask at the outset how they usually handle disagreements and the unexpected, because every job includes surprises, no matter how well you prepare.

If a builder isn't listening at your first meeting, the chances of them listening in the midst of the build are slim. So move on and look for someone who wants to work collaboratively. Then you'll be in with a real chance of achieving your beautiful home.

And there are some amazing builders out there. I heard recently of builders who were told by their client: "Thank you – you made it so easy." And that was after the client had lived for four months in a caravan in the garden! So it is definitely possible.

Making the decision

If you've done your preparation well and you believe you could have a good working relationship then the deciding factor is often cost. This is such a big issue that I've devoted a chapter to it (see chapter 7).

They will want tea and more tea

I once asked Giorge, a local builder, what he looked for in a good customer. His answer came very quickly: "Someone who keeps the lads supplied with tea, makes quick decisions and pays on time."

We did our retrofit during Covid and my way of switching off from the scare, frustration and building mess was to cook. Picture this – we had just one tiny area of work surface left. I had my hob and the ovens were still in place. Otherwise my kitchen was a building site with no windows or doors to protect us from the autumn and winter weather. Never mind aprons – I was in my thick coat, boots and a hat.

Not only did I cook, I followed online lessons with Bread Ahead*. We were living in the worst mess, cold and mucky, but we ate well! So many different cakes, breads and biscuits. I'm sure part of the good relationship we built with our builder was the food. We had the best-fed brickies in the country!

THE BUILDER WON'T TELL YOU THIS

There are some situations builders live with every day, but won't think to tell you. It's second nature to them so they assume you know too:

You will be living with these people: if you decide to live in, then you'll wake up each morning to these faces. They'll see you in your dressing gown, when you're on top form and when you're down in the dumps. Although this is your home, for them it is a place of work. They may be really good at caring for your space or they may not notice the 'home' bits because they're focused on the building bits.

To manage this, make yourself a retreat in the house. Having somewhere to get away is really helpful. This is easy if you're working on a small part of the home. If you're doing a major piece of work then it may just be your bedroom – or at least where you sleep. Make clear this is out of bounds – that you want them to knock or shout if they need to speak to you when you are in there.

If the work is all-encompassing, you may decide to move out. This can be more comfortable and definitely cleaner. But it also costs money – money that could go into the renovation – so you need to weigh it up carefully. If you live in, good builders will allocate time to tidy up at the end of each day. If you're elsewhere, they can use this time to carry on working.

One benefit of living in – if you can stand the mess – is that you can keep a much closer eye on what's being done. You can check at the end of each day and ask for adaptations and changes as you go along. The once-weekly visit when you live out can make this more challenging. So it depends how much of a control freak you are, how well you understand building and how much you want to be involved.

Builders don't know which bits of your home matter most: remember this is their workplace and they do a mucky, dusty job. If you want your house to be cared for, you'll need to take some responsibility. You can't expect it as automatic.

We always provide dust sheets. Any old sheets will do. It's like tripping down memory lane for us when the builders come in – we've still got the old duvets from when the kids were little. They do a great job of protecting carpets!

You'll find that some builders bring their own dustsheets and are really careful. Others just don't bother – I think it doesn't occur to them. When we put dust sheets on the stairs, we had to adjust them each time we went down or up. Not sure the guys ever noticed.

It's OK to point out when things get a little messy – they might just need a bit more prompting.

Then we've had builders who were so careful and cleaned up after themselves. When Harry and Jack worked with us, they were better at cleaning up than me!

So be ready to set a tone at the outset. Dust sheets out, precious items stored safely out of harm's way, furniture covered up, kettle and supplies in easy reach of the tap. Find out beforehand what to expect: who is going to be there and what is their routine – i.e. what time do they arrive/leave etc.? Then you can be as prepared as possible.

You won't always know when to expect them: the building day generally goes from 8am to 4pm. Some builders work five days a week or use Saturday as a time to catch up if they've had to miss time during the week. So be prepared – you may only have one day a week off.

Something I didn't understand for a long time is the juggling builders have to do between customers. If you choose to work with a medium-size or small building firm, they may have more than one job going on at the same time.

Their challenge is to make sure the people they employ are always working. They don't want:

- a carpenter waiting to put up cupboards because the plumber hasn't finished putting in the sink
- the floor fitter waiting for the underfloor heating to be finished
- the brickie having to wait for the scaffolding to go up.

Every delay of this nature costs money.

Sadly this probably means downtime for your project. You might be all prepped and ready at 8am to have no one turn up. They may come later in the day but more likely you won't see them until the next day or next week. If this happens a lot, it gets really frustrating.

Hence the need for you to make quick decisions – or at least be ready with your decisions.

- Know which tiles you want in the kitchen.
- Decide how many power sockets you want and where you want them to go.
- Say which bathroom suite you want – especially if it has to be ordered ahead of time.

Leave decisions to the last minute and you risk holding up the process. Your builder will have allocated a day for the electrician or tiler to attend, but if you are still making up your mind, the work can't go ahead. It will still have to be paid for, unless the builder has other work that person can do instead.

I imagine you won't like hold-ups like these – I hate them. I can stand the mess as long as we're moving forward, so downtime drives me up the wall. Delay not only costs time, it also costs money. If trades turn up and can't work, the cost of their time will end up on your bill at some point.

Ask at the outset how many jobs they have going on. Find out how likely it is that you'll have days when nothing is happening. And if it is likely, agree how they'll let you know.

And, of course, ask when particular decisions need to be made. Be clear how much warning you need to shift from endless browsing on Pinterest to putting in your order and arranging delivery.

Parking: Unless you live on a grand estate, you'll have to manage parking. Builders will turn up with at least one van and probably more.

- If you have a drive that helps, but remember you'll probably need room for a skip as well.
- Check out parking permits - does your council provide specific permits for builders or can they park for free when working on your house. It's unlikely, but worth checking.
- The builder may offer to pay 'themselves'. This will end up on your bill, so find out what they charge. I've read about a builder who charged £70 per day to cover the cost of a fine should he get one. That way he could just park anywhere and save time!

The van will hold tools, materials, personal belongings. It will also be a quiet place to go for a break if needed, so it's much better near the house. Parked down the road, they'll spend time trooping backwards and forwards – it's inconvenient and takes time away from the work.

Deliveries: on the subject of parking, it's helpful to know what to expect with deliveries.

Stroppy delivery drivers

Some delivery drivers are delightful and helpful; others seem unimpressed with the world and certainly aren't enjoying their jobs very much.

We live at the end of a cul-de-sac so I can see we're a trial when it comes to delivering big items. Most drive up the road quite happily to drop off the delivery. Some manage it with a grumble, others refuse to even start.

Of course the most stroppy one we've ever had was delivering the materials that were needed immediately or work would be held up. He walked the length of the road to tell us he wasn't prepared to drive up. It took a lot of persuading to change his mind.

Obviously, he made it – it's a challenge but so many others have done it without quibble – but you'd think he'd crossed the desert, so ripe was his vitriol. I managed not to point out that the crane driver had been up the day before to deliver massive steels! Just to make himself feel better, he dumped the pallets of insulation on the road just where the neighbours would need to park their car.

I do understand and I do appreciate the effort required to get materials to us in good time. The challenge comes when the driver implies we bought that house just so we could be a nuisance on that particular Thursday.

But it's par for the course and part of the rich tapestry of renovation and retrofit!

If your location has any quirks or restrictions that could affect the movement of large vehicles, be clear with the builder about the best way to manage it. For example, some materials will be delivered on a pallet – timber, insulation batts or bricks will be loaded onto a wooden pallet that can be lifted up in one piece by a forklift truck. The pallet makes it easier to load the lorry, but it may not make it easier to unload outside your house.

When you have plenty of room around your house this will be inconsequential. If you live in terraced housing or on a narrow road, this is a constant hassle that needs working out ahead of time. The last thing you want is for your materials to be driven away again in a fit of pique.

When a pallet is delivered, it is likely to be on a 'kerbside only' basis. This means the materials are dropped off on the pavement in front of your house. It will be for you or the builder to move them onto your drive, into your garden or just out of the way of neighbours.

We've found ourselves loading bags of cement onto a wheelbarrow to shift it into the garden after it was delivered on a Friday afternoon, moving huge packs of insulation that needed not to get wet and shovelling rubble till we were exhausted.

Neighbours: which brings us to the thorny question of neighbours.

If your neighbours have done work like this themselves then they are more likely to be understanding. On the other hand if you live in a road where loads of people are renovating, building fatigue may be setting in. It's irritating to be held up by builders' vans when you have a train to catch. Or if those noisy deliveries turn up just as the baby goes to sleep.

So we need to take care of each other. Your neighbours will be there long after the builder has gone so it pays to be considerate.

- Let them know when deliveries are expected or when the crane is coming to bring in the steels.
- Apologise in advance and talk them through the plans so they know what to expect.
- When noise or traffic has been really bad, take flowers or a bottle of wine to thank them for their understanding.

It's bad enough living through building work when you know you'll eventually end up with a lovely home for your labours. Neighbours have to put up with the noise and disturbance with nothing to gain. And then you have to live alongside each other, so be kind and caring. Your work will finish one day and then soon you'll be on the receiving end of someone else's ambitions.

While the builders might be an irritant to the neighbours, it's also possible that the neighbours could be a thorn in the side of the builders. So always let the builders know if someone is particularly tricky so they can be aware and work around it. Hassle takes time, so the more you can avoid it the better.

Want to get your hands dirty?

If you want to get involved and do some of the work yourself – just for the joy of it or to save money – then best to agree this with the builder ahead of time. They may have concerns:

- They'll have to incorporate your efforts into their work so they'll be concerned about what sort of job you'll do.
- They have to answer to the building inspector, so need to be sure you'll do a good job.
- If you don't work to the same timetable you could hold up the wider process.

It's a great idea to take part in the build, but make sure you're really committed. If you suddenly become too busy or realise the job's more than you thought, the builder will have to pick up for you. That means allocating time in the schedule and finding trades who can fill in at short notice.

Most important, recognise what you can and can't do. If you're going to have to search on YouTube to work out how to do it (unless you're particularly good at learning new skills) – maybe leave it to the professionals. If you really want to help out, then taking on the tidying can leave the builder to do the specialist work.

We can do that ourselves

It was one of those late-night decisions – we would insulate under the suspended floor ourselves. Part desire to be involved, part concern the builders wouldn't be precise enough, part not sure how to explain what had to be done.

I now know that Ecological Building Systems* have a great blog and all the materials to do the job easily, but at that point John was having to work it out in his own head. So it made sense to start and work out the best way as he went along. (His solution turned out to be the same as EBS's, so he was very proud of that.)

We started full of vigour, then life took over and we lost track. Not helped by the plumber who was busy elsewhere and gave no idea when he'd be in to put down the underfloor heating.

And of course, we suddenly got a message to say he'd be in the next day. And of course we weren't ready. That was one long night, but we got it done. And of course, the plumber let us know next morning that he would now be in on Monday.

So be sure you have the appetite for a consistent bit of hard labour. And if not, hand it back to the builder with plenty of notice.

Summary

- Finding a good builder is a challenge so it's worth taking time to get to know possible providers.
- When problems arise it's often about money. Keep notes of any extra spend and be prepared to have tough conversations if needed.
- You'll be living close to the builder for the duration of the build, so do your best to make it a positive relationship. Remember it's their workplace. It's up to you to protect the parts of the house that matter most to you.
- Sort out parking and delivery – it's the small things that make a big difference.
- If you want to get involved, agree it with the builder. You may have to reach the standard they set if you're doing work that involves the building inspector.

To find the information indicated by the * follow the QR code.

NOTES

CHAPTER 6
CONVENTIONAL MATERIALS THE BUILDER WILL USE

I love this stuff – so get ready – I'm about to get my geek on!

The health and energy efficiency of your home will rise or fall by the materials you use. Sounds very dramatic, but I've been amazed at what I've learned since we worked on our home and I've settled into my new role as 'building bore'.

A lot of the information in this chapter won't be known to your builder. It's up to you how much you want to communicate, but do make sure you get the quality of home you're looking for.

There are a number of reasons why the materials you use are so important:

- volatile organic compounds (VOCs)
- embodied carbon
- suitability for your home.

VOLATILE ORGANIC COMPOUNDS

VOCs are found both in nature and in numerous products we use every day in the home. Some are harmless and some cause real difficulties, because they evaporate easily into the air. In essence they pollute the air in our homes. And since we spend over 80% of our lives in the home, this is a real problem.

They are a group of carbon-based chemicals that evaporate at room temperature, releasing gases into the air. Common examples of VOCs that may be present in our daily lives are: benzene, ethylene glycol, formaldehyde, methylene chloride, tetrachloroethylene, toluene and xylene. Overall, they're pollutants that cause a number of health problems such as headaches, ear, nose and throat irritation, and nausea. They can also have long-term effects on kidneys, liver and the nervous system.

The biggest clue that VOCs are in the building is the strength of the smell:

- that new-paint smell we like so much – VOCs
- the smell of bleach or cream cleanser that we associate with a newly cleaned home – VOCs
- the new-car or new-sofa smell that is so exciting – VOCs.

VOCs are found in a wide range of products like paints, cleaning supplies, fuels and, most relevant to us in this chapter, building materials. They are also released from natural sources like plants and mould*.

VOCs in building materials

The more synthetic the material, the more VOCs will be involved. In many cases the primary risk is when they are first installed, because this is when they'll off-gas most – off-gassing being the process of releasing toxins into the air. Off-gassing can last anything from a few days to six months and, in some cases, even years.

As a homeowner you may be able to get out of the way during the installation work and avoid that early risk. This is also a tick in the box for considering moving out while the work is done. Not such good news for the builders, though. They're living with a constant barrage of VOCs each day they're at work and I'm not sure how many of them know this.

VOCs are released from insulation, adhesives, paint, plaster, expanding foam, composite wood products, varnishes, flooring – the list goes on. There is a fascinating Grand Designs programme about a family who retrofitted their home without VOC's because of their children's asthma* Worth a watch if you have respiratory problems.

Natural materials may have VOCs but they'll be in much lower concentrations and less harmful. As soon as chemicals are actively involved, the risks increase. So for this reason alone, there is an advantage in using natural materials.

EMBODIED CARBON

Embodied carbon refers to the greenhouse gas emissions released through the entire life cycle of a product, taking into account:

- raw material extraction
- manufacture
- transportation
- construction
- maintenance
- end of life – how the material is disposed of.

If you have concerns about the climate, then considering the embodied carbon of a material will soon become second nature. Looking at the 'whole life' of a product requires you to understand where it was produced or grown, what is involved in turning it into the actual building material, the impact of transportation and what happens to it when it's no longer useful.

SUITABILITY FOR YOUR HOME

When it comes to improving the energy efficiency of your home the focus has always been on keeping warm in winter. Government schemes focus on insulation and high bills in wintertime, so it's easy to assume that this is all we need to concern ourselves with and that one insulation is as good as another.

It's also easy to assume the builder will know what's needed and have information about different materials at their fingertips. Or that they'll follow the architects' instructions about choice of insulation. Sadly the decision about the materials you have in your house is more likely to be driven by what the builders' merchant has in stock and which is most familiar to your builder.

This is why I suggest you get involved if you want the best outcome for your home. Unless you can find a specialist eco builder – maybe a heritage builder or someone who's trained in retrofit – you'll end up with a synthetic material with high Lamda value, high embodied carbon and short decrement delay, such as: a foil-backed petrochemical board. (Don't worry, explanations of these terms are coming!)

The end result? Warm in winter with plenty of condensation and little protection from the increasing heat of summer.

What you actually need is a material that will keep your home at a steady state in both the coldest winter and the hottest summer. To achieve both you need a material that has:

- a low Lamba value that slows the loss of energy through your walls, floor and roof in the winter
- a long decrement delay that moves heat from outside into your home very slowly in the summer.

I fully expect that last paragraph to be a step too far. Even I have to stop and check all the terms on a regular basis. But fret not – here's all you need to know about insulation terminology:

INSULATION-SPEAK

This is like learning a whole new language. You don't need to know detail unless you like getting technical – in which case you'll need to go searching, because I don't! The reason to name all the different measures is that some people use one lot, some another.

This part of the chapter is worth a quick skim and then come back to it when you're actually looking for the most suitable insulation. I'll do my very best to keep this short and sweet so you know in simple terms what to look for. Then we can move onto the exciting stuff!

Why bother with the different measures?

The success of your home improvement project will depend in large part on the materials you use. And if you're going through all that pain you want to know that you'll be warm, comfortable and better off at the end of the process.

Best-case scenario: your builders would have the same ambition. However, while you want to be warm in winter and cool in summer, your builder may be focused on compliance with building regulations and getting a pass from the building inspector. For that reason you need to stay on top of this and make sure you are getting what you asked for.

The measures you need to know are:

U value (thermal transmittance)

- Measures how well a system (i.e. the different layers of a construction) transmits heat.
- Measured in watts per metre squared per Kelvin (W/m^2K).
- **Lower the better** – you want heat to stay where you put it.

For example, a single pane of glass has a U value of $6.0W/m^2K$ but a triple-glazed window has a U value of $0.7W/m^2K$ and that's a huge difference to your energy efficiency if you have a lot of windows.

Thermal conductivity/K value/Lamba value

- A measure of the ease with which heat can travel through a material by conduction; this is intrinsic to that material.
- Measured in watts per metre Kelvin (W/mK).
- **Lower the better** – you don't want heat to flow.

R value or thermal resistance

- Measures how well a material resists the flow of heat. It links the thermal conductivity or Lamda value to the thickness of the material, so the greater the amount of material the less heat flow.
- Measured in metres square Kelvin per watt (m^2K/W).
- **Higher the better** – you want it to resist the heat transfer.

NOW FOR A DEEPER DIVE

So far so good – now you can be warm in winter. But as the climate changes we increasingly have to think about being cool in summer, too.

Aberdovey holidays

Now I'm really going back in time!

When I was a youngster my parents had a very basic old caravan that we towed laboriously each year to Aberdovey. I think it was made of plywood or cardboard!

My main memory is of being freezing cold and damp, as the rain lashed down outside the window. Once a week there was a film in Towyn and we would go – whatever the film – just to get away from the rain. Reasons for seeing a Tarzan movie must be many – staying dry was mine.

On rare occasions there would be hot sunny days. Still that horrible caravan provided only meagre protection – this time leaving us sweaty and bad tempered. The van was overheated by breakfast time and Towyn still looked appealing.

Got to say – I've never been to Aberdovey since. And I'm fanatical about my home providing protection, whatever the weather!

Insulation is not only there to keep us warm in the winter – it can also keep us cooler in the summer. Not many builders are well informed about this aspect of insulation, so it's down to you. There is so much we can do now to protect ourselves from overheating just by using different material properties and understanding how materials handle heat.

Thermal diffusivity

This tongue twister is a measure of how well a material holds onto heat, then releases it. This is its buffering ability. Thermal diffusivity tells us how we can reduce the impact of very hot summer days on the internal temperature, especially in those hot attic spaces and sunny rooms.

Three factors contribute to thermal diffusivity, affecting how a material handles heat:

Specific heat capacity

- The amount of heat energy required to raise the temperature of a material by a certain amount.
- Indicates how much thermal energy a substance or material can store.
- Measured in Joules per degree Celsius (J/degree Celsius) or Joules per Kelvin (J/K).

Thermal conductivity - see above

High thermal conductivity transfers energy quickly – e.g. copper has a Lamdba of 401W/mK, which is why it's useful on the bottom of our kitchen pans

Density

- Density refers to the mass of a material per unit volume.
- It links directly to thermal mass, which is the measure of a material's capacity to absorb, store and release heat over time.

High specific heat capacity + low thermal conductivity + high density = thermal diffusivity

Wood fibre and hemp are great examples of thermal diffusivity. They resist the transfer of heat into a room, meaning the heat outside stays there while you remain cool in your home.

Thermal mass

- The ability of a material to absorb, store and release energy.
- High thermal mass means materials take a long time to warm up and cool down.
- High thermal mass helps to smooth out extremes of temperature, maintaining a comfortable internal environment and reducing the need for heating.

Wood fibre and hemp will absorb the heat from outside and hold onto it for nine hours or more, then slowly release it into your home as the day cools. It's this thermal mass that will make sure your home stays at a steady temperature whatever the weather.

CONVENTIONAL MATERIALS THE BUILDER WILL USE

Decrement delay

Great label – a great one to use if you want to be impressive!

Summers are definitely getting hotter, so if we want our homes to be fit for the future.we have to build in protection. When the heat really hits, the last thing we want is for the internal temperature to be the same as the external temperature. That's a recipe for frayed tempers on a monumental scale!

Decrement delay or time lag describes the time it takes for the peak heat of the midday sun hitting an outside wall or roof to get into your home:

This time lag for some materials can be very rapid (e.g. in a steel-framed house) so the heat reaches the inside quickly, but in other houses (e.g. a stone cottage) the heat takes much longer to reach the inside so it feels cooler internally. Insulation materials can also affect the heat transfer rate and a decrement delay of between eight and 12 hours is what we're looking for.

External temperatures can vary from around 10°C to 30°C in summer but the right materials mean the walls can buffer the wide swings, so we need never know what the weather is like outside.

This ability to reduce the swings in temperature is known as the decrement delay. The longer this is, the better the building fabric is at reducing these swings.

Longer the better – you want a long decrement delay, so it will take many hours for the heat to be transferred.

Just before we move into the real stuff of insulation, better check out who this guy Kelvin is!

The Kelvin scale is similar to the Celsius scale or Fahrenheit but is based on absolute zero – the coldest possible measurement. There is no such thing as a negative temperature in the Kelvin scale, so it enables scientists to be more precise.

I looked up who created it and it was William Thomson. Made no sense at all until I saw he was also Lord Kelvin! So thank you, M'Lord. Looks like you made a lot of things simpler – as long as we can all just recall whether we're looking for higher or lower the better!

BASIC MATERIALS

Concrete is the second most used substance (after water), the most widely used building material and the most manufactured material in the world. Over the past 100 years it's revolutionised the built environment, providing us with homes, industry, roads – most things you can name in our modern life. Over 14 billion cubic metres has been produced. So it's easy to see why builders will default to this when digging foundations.

Cement is a binding agent, typically a fine powder, that, when mixed with water, creates a paste that hardens and binds other materials together. It is a major ingredient in concrete manufacture, which is why the two constantly get confused. In addition, it's used to bind other materials together, the most obvious being – in the form of mortar – building brick walls. It's part of screeds and renders, can be used to fill cracks and repair damage, even mixed with soil to improve strength and stability.

Cement is made from limestone or chalk, clay or shale, which is heated and ground down to form the powder we are familiar with. Together with the concrete produced, this accounts for 8% of global CO_2 emissions because it needs high temperatures in manufacture and billions of cubic metres of water.

Modern brick is made from clay and sand, then fired to 1,000°C, which makes it a durable building material. At that temperature, there are going to be environmental implications – the energy used to take the bricks to such a high temperature, water needed in the process, dust pollution etc.

All three of these materials are familiar, reliable and easily accessible, so clearly will be used by builders who probably don't know of alternatives and can see no reason to deviate from what is known and works

See chapter 11 for alternatives.

INSULATION MATERIALS

Most builders will default to the familiar – honestly, don't we all! Unfortunately, familiar when it comes to insulation materials tends to focus on the synthetic.

At best, your builder will discuss the options with you and recommend which one they believe is best. At worst, they'll just go ahead and get ordering, just as they won't check with you that you want bricks for your new extension.

The main options they'll put before you are:

- PIR insulation/PUR
- fibre glass/glass wool
- XPS/EPS
- mineral wool
- spray foam.

Read the following, then talk with the builder about any concerns you might have. You may find you're telling them about issues they're not familiar with. And they may have other views to give you.

Rest assured if you want more information there is loads available on the web. Use the terminology you have here and go searching. Most important, don't just take the word of anyone else, look into it all first and make an informed decision.

NB: check very carefully who is responsible for a website you're looking at. I've done so much of this and find glowing reports of materials I know to be toxic, only to discover the manufacturer runs the site. It's all in the fine print, so keep looking and cross-checking.

Wrapping in plastic

Synthetic insulation products are built around petroleum, so a form of plastic. If you do decide to go with these in your home, you <u>must</u> add ventilation. In effect you will be wrapping your home in a big plastic bag, which means there'll be nowhere for vapour and moisture to go.

Upgrading with any insulation will require ventilation, but the use of plastic based materials makes this imperative. So please don't miss this or you'll have other problems to face even though you'll be warm as you sort it. (See chapter 10)

PIR/PUR insulation

PIR is the generic name for the most common synthetic insulation material. I can almost guarantee the builder will assume this is what will be used.

It is made from a chemical reaction between polyop and isocyanate which forms a type of rigid foam. It does make for good thermal insulation – the builder is right about that. You'd recognise it from the foil facings on either side. The foil is there to improve heat retention but it also makes it easily recognisable. Most of us have seen it leaning up against a skip when the builders are in.

There are health risks when using it – primarily for the builder:

- sensitises the respiratory system, leading to allergic reactions, asthma and breathing difficulties
- skin irritation – direct contact can cause dermatitis and other skin problems
- eye and throat irritation.

As a result, builders should only work with it when wearing personal protective equipment (PPE) because of the dust produced when cutting it. Can't say I've ever seen this!

- **VOCs:** there is some off-gassing, but it's considered low risk. The biggest problem caused by the off-gassing is that the insulation shrinks slightly, leaving small gaps, which affects the efficiency of your system. Which means you're not getting the quality insulation you thought you were.
- **Embodied carbon:** made from petrochemicals, releasing loads of carbon in its manufacture. No way is it eco-friendly.
- **Suitability for your home:** you can use this if your home was built after 1940 and has a cavity wall, since the materials used to build your home are not vapour permeable (breathable). However, be aware that you're wrapping your home in plastic, so you must make sure you have good ventilation if you want to avoid condensation and potentially mould.

It is good for winter and will help your home be a lot warmer. However, it has a short decrement delay so will be next to no help in summer.

It is not good at managing moisture. Just think of times when you've seen it sitting in puddles: it's not affected – it hardly notices the puddle is there. That may sound like a benefit, but if you have condensation in your room, it has nowhere to go – all it can do is sit on your wall and wait for the mould to grow.

Also be aware:

When burned, PIR releases toxic fumes including hydrogen cyanide and carbon monoxide. Both gases are colourless and odourless so can incapacitate individuals before they're able to escape.

Fibre glass/glass wool

Loads of people already have this in their homes – often in the loft, ceilings and walls. You'll see it in pink rolls looking a bit like candy floss. It's easily available and a good price so often a first port of call.

It's made from recycled glass, which in turn is made from sand, so a renewable source. The glass fibres create plenty of air pockets and, since air is a poor conductor of heat, this means the heat stays trapped in the material.

- **VOCs:** formaldehyde is used as a binder and this will off-gas, potentially causing respiratory irritation, headaches and other health issues. However, you can now get formaldehyde-free versions, so ask your builder to look into this if you decide to use fibre glass.
- **Embodied carbon:** production of fibre glass is very energy-intensive because of the heat needed to melt glass, which gives it high embodied carbon.

- **Suitability for your home:** it's OK as an insulator, but low mass (it's light and airy) so not so good at keeping you cool in summer. It is good for acoustic insulation.

Good in some ways but not ideal overall, so if your builder suggests this it's worth asking for the specific reasons why. Then look into it yourself to see if you agree.

XPS (extruded polystyrene)

XPS or extruded polystyrene is a rigid foam insulation made from polystyrene. It's made by heating polystyrene resin beads to create a closed cell structure which gives it a density, strength and moisture resistance that are particularly useful in some circumstances.

- **VOCs:** not a problem once it has been installed but some VOCs are present in manufacture because it requires blowing agents to get the foam structure. Special care also needs to be taken if the material is exposed to high temperatures. So builders using a heat gun near it should always take protective measures. It's possible now to find XPS with lower or zero VOCs, so ask your builder to look into that if it's the insulation of choice.
- **Embodied carbon:** has a higher embodied carbon because the blowing agents used to get the foam structure contain hydrofluorocarbons, which are potent greenhouse gases. There are versions now made using lower-GWP* (global warming potential) blowing agents, such as HFOs (hydrofluoroolefins), to reduce the embodied carbon of XPS. Nevertheless, it still has significantly higher embodied carbon than materials like cellulose, wood fibre and even some other foam insulations.

- **Suitability for your home:** lightweight, easy to get hold of; resistant to water and vapour, so good in damp environments; excellent compressive strength so good over foundations and floor slabs; not vapour permeable so not suitable for an older house. As with PIR you are wrapping your home in plastic so good ventilation is a must to save you from condensation issues.

Also be aware:

It will expand, melt and burn at high temperatures, releasing harmful gases.

Mineral wool

Mineral wool is popular, particularly in commercial buildings because of its fire-resistant qualities. It's a rock-based mineral fibre made from basalt rock and recycled slag. The minerals are melted and spun into fibres.

- **VOCs:** made with inorganic materials but can contain trace amounts of VOCs due to the binder used during production. However, modern mineral wool products often use formaldehyde-free binders, and all products are tested to ensure VOC emissions remain below established limits.
- **Embodied carbon:** the manufacturing of rockwool involves melting basalt rock at high temperatures. It also contains slag that's a molten byproduct of steel manufacturing – all of which makes for an energy-intensive process. This can be offset to a degree by the fact that it will last a very long time.
- **Suitability for your home:** excellent fire resistance – it won't ignite or contribute to the spread of fire; good soundproofing between rooms or

outside; good thermal resistance, so keeps the building warm in winter and cool in summer; it is vapour permeable, but it's not capillary active – i.e.: it can't wick the moisture away, so it just continues to get wet.

Also be aware:

It needs protective eye protection, gloves and masks when handling.

Spray foam

Think <u>very</u> carefully before ever installing spray foam.

One of the biggest problems with this material is the amount of hard selling involved. Cold phone calls can often be agents attempting to sell to people who don't understand what they're signing up for.

It is made from polyol resin blend and isocynante. When combined, they react and create a polyurethane foam that can be used as insulation.

- **VOCs:** releases VOCs during and after application, including formaldehyde, acetone, toluene and others that can lead to throat irritation, coughing, shortness of breath, skin and eye irritation, headaches and nausea. Off-gassing reduces significantly after the foam cures, but it is still best to make sure there is good ventilation around it. Spray foam dust, consisting of small particles and fragments of the cured foam, can be inhaled and cause respiratory irritation.
- **Embodied carbon:** spray foams use hydrofluorocarbons as blowing agents to create the foam structure. These have a high global warming

potential (GWP). There are now newer versions that use water or HFOs (hydrofluoroolefins) as blowing agents, which have significantly lower GWP.
- **Suitability for your home:** good insulator and creates an airtight seal, reducing draughts; good soundproofing. **BUT** makes it difficult to get a mortgage: some insurance companies won't insure a home with it in; impossible to remove easily; vapour-impermeable so encourages damp, condensation and mould.

The last three points are worth reiterating. Once spray foam is in your home it can be almost impossible to remove.

Also be aware:

Check with your insurance company before making any decision. The problem is that moisture can get trapped behind the foam and you won't know anything about it until you have a massive problem on your hands. It's such a thorough covering that it's always impossible to know what's going on behind it.

The dangers of spray foam

It was quite common at one time to get a phone call from a salesperson telling you how wonderful it would be to have spray foam in your home. You'd be warm, reduced bills, increased home value – all the attributes of a well-insulated home.

However, it's since been discovered that it's really hard to get insurance on the property and when the time comes to sell, mortgage lenders aren't keen. There are just too many concerns.

Once in place, it can only be removed at great expense. I heard of one person who wanted to sell their home, but any potential buyers pulled back as soon as they spoke to mortgage lenders. The only option seemed to be taking out the offending materials.

Believe it or not, the most efficient and cost-effective way was to remove the roof itself because it wasn't possible to get the foam out and leave the roof intact.

So if you do think it could be a good idea for your home, please look into it very thoroughly. There's plenty on YouTube to choose from. I'd hate you to get stuck with a home you can't sell.

WHAT GOES OVER INSULATION?

Once the insulation is in place, there is still a job to do. You need a smooth surface to put plaster on, so something is needed to cover up the insulation.

Plasterboard: one of the most commonly used materials is plasterboard. This is a panel composed of gypsum (calcium sulfate dihydrate). It's also called drywall, wallboard and gypsum board. It's mined, crushed and heated to make it into what we know as plaster of Paris. Which brings back all manner of childhood memories!

The plaster of Paris is mixed with water and additives, laid out into board and dried. Then it is cut into the shapes most needed by builders. It can be mashed up and remade into more board, added into cement production or used as a soil conditioner in agriculture.

So in theory it shouldn't be too bad from an environmental point of view. Yes, it needs mining, but it is eminently recyclable. However, having seen mountains of old bits of plasterboard in the tip, I am yet to be convinced that this actually happens. It can be 100% recycled so in theory is pretty green, but it has to be disposed of separately because of 'potential reactions'.

Ordering plasterboard will be automatic for your builder. It is probably the basis of every smooth surface you see around your house right now. Its job is to cover the insulation or wall and it can be used with all the materials referred to above. The reason it is so important is in the title – it provides a smooth surface for plaster to be applied.

Suitability for your home: The main issue is breathability. It is vapour permeable, as long as it is used correctly – and can be part of a wall build-up with wood fibre. You just need expert advice to make sure it is right for your home

Plaster: this is the gloop that goes on top of the plasterboard to give you a smooth surface to paint on. There are different versions but the most commonly used is gypsum plaster. Made with a mix of gypsum plaster powder, water and sand. It's an art form to put on plaster well:

- a thin skim on plasterboard – usually a 3mm gypsum coat done in two layers, usually dry in about one week ready for painting
- if going over bricks, a thicker coat is applied – of at least 10mm – but this is rare these days
- a mesh embedded into the skim to hold the next layer or a scrim tape over the plasterboard edges and angle beads to stop cracking.

Environmentally, the same issues apply as for the insulation. The plaster is made with the help of binders, which can release VOCs, especially in manufacture and when being applied. There are other forms of plaster that are much more eco-friendly like clay and lime. (See chapter 11).

CONVENTIONAL MATERIALS THE BUILDER WILL USE

In terms of impact on family members, plaster obviously has a lot of moisture to get rid of as it dries and this can be trying to live with.

Paint: this the final layer in your wall build-up, and often the most exciting. Now you're finally getting to create the beautiful home you've been looking forward to. Endless poring over colours can be totally absorbing.

Sadly there are also issues with paint when it comes to the environment, healthy home and the safety of your home. This is relevant for two reasons:

- standard paint off-gases – that new-paint smell we have all come to love is actually the smell of toxins as the paint dries
- many paints contain plastics, so are not vapour permeable. The end result is that moisture can be held behind the paint until it eventually peels off. All of which results in damage to the wall or brick behind the paint*.

Off-gassing: you can find low-VOC or no-VOC paints. Typically they're measured in grams per litre (g/l) and there should be information on the tin or the website of the manufacturer. You can also look on Green Seal* to check on the brand you're leaning towards.

Breathability: To check if a paint is suitable for a vapour-permeable/breathable building, you need to check the SD value. This stands for steam diffusion and indicates how easily water vapour can pass through the paint coating. A low SD level means the paint is more breathable, so water vapour can escape easily. A truly breathable paint will have an SD value of one or below.

Most paints are breathable to some degree, except for vinyl. Hence the importance of the SD value, because any question about the breathability of the paint will receive a 'Yes' answer. But it may not mean that it's as breathable as you need for your old home.

NB: also check on the fillers used, as these are often full of plastic, although you can obtain ones that are compatible with lime plaster. It just needs a bit of thought and effort to work it out.

IT'S A PRETTY BLEAK PICTURE

So many of the products in everyday use are synthetic and they have a significant impact not only on the environment, but on our homes and our families.

The big sadness is that you probably won't be told any of this. Not through any form of malice or forethought. Just because people operate from habit and have stopped even wondering about alternatives. In fact, most builders won't even know alternatives exist.

But I'm here to tell you there are some fabulous alternatives out there. All you have to do is read on to have your mind blown. It's exciting!

SUMMARY

- You can assume builders will use the materials they are familiar with – and that are easy to obtain – whether they suit your home or not.
- There are measures that tell you about the efficiency of insulation materials. Use the list above when you are shopping around, because different brands will use different measures.

CONVENTIONAL MATERIALS THE BUILDER WILL USE

- The usual building materials are oil-based and made with petrochemicals. Be informed about the risks and benefits of them, because your builders probably won't know.
- Some of the materials suggested will make it difficult to get a mortgage or sell your home in the future.

To find the information indicated by the * follow the QR code.

NOTES

CHAPTER 7
WHAT YOU NEED TO KNOW ABOUT COSTING

This is the million-dollar question. And we all know the analogy of a 'piece of string'.

It's so easy to get carried away. Fuelled by Pinterest and social media generally, we have hopes and dreams that keep us going through the challenges of prepping for a retrofit and renovation.

First talks with a builder can be like a bucket of cold water. 'I had no idea it was going to cost that much!' And why would you? This is you entering a new world and you have to adjust your thinking to the harsh reality.

But there is some information that can help before you take your first steps.

BE PREPARED

By asking for an estimate you're giving the builder a lot of work – it takes time to put it all together. They have to work out staffing, material costs, obstacles, the time frame and how it would fit with other work, incidentals – the list goes on. So don't be surprised if some don't even reply or send a 'back of a fag packet' summary of cost.

An estimate is a non-binding 'guesstimate' of the potential costs and can be worked out on the plans you submit for planning permission. It will give you a guide price. Just remember it's a guess and don't set your heart on it.

A quote is a legally binding calculation of the actual costs. To get this you need to share the drawings prepared for building regulations – they contain a lot more construction details and are the plans the builder will actually build to. The quote is a number you can set your heart on, but do make sure to find out the time frame for the quote. You can't return to a quote months later and say you want to go ahead, expecting it to be the same – prices may well have changed by then.

The builder will need detail ranging from the big picture to any specifics you're looking for – everything from:

- what you want the end result to look like
- how many power sockets you want and where
- specific materials you want used – natural insulation materials versus synthetic
- elements of the building you don't want touched
- whether you'll live in or move out.

Depending what you're asking for - a quote for a watertight shell or a quote to cover everything including painting and decorating - you'll need to provide the appropriate detail.

This isn't easy – it's a big job for you, too. It requires you to imagine what you want (which I'm hopeless at) and to be able to describe it effectively. Of course, a good builder should be able to help you sort out the detail, but I'd recommend you don't rely on that.

WHAT YOU NEED TO KNOW ABOUT COSTING

Before you start meeting professionals about your build, do your homework:

- Get a Pinterest board going or look at home magazines and save pictures that show what you want.
- Imagine yourself living in the place – where would you sit, where would you want a light or a plug for your phone?
- Visit your friends and find out what they love or wish they'd done differently.
- Visit loads of showrooms to explore options for kitchens, bathrooms, bedrooms, furniture.
- Draw up your own plan and include cutouts of important items so you can move them around to see what works. Or if you're techie, find an app to do it for you.

Above all, talk to people about your vision, then get them to play back what they heard. How close is their description to your image? Ask them to describe in detail so you find out what landed and what didn't. It sounds like a bit of a kerfuffle, but it will be worth it in the long run.

The art of communication

Before my sudden shift into retrofit, I worked in Leadership and Management Development. I know! An eye-watering change.

I spent a lot of my time helping leaders communicate clearly with their people, so I can confirm that it is one of the most difficult tasks. There are just so many ways it can go wrong – and you'll probably never know it.

You talk to your potential builder about the work you want done and how you want your house to look. In your head there is a point of reference – a picture, a list of requirements, a house you've seen – and as you describe your requirements, you assume the builder has the same point of reference and sees what you are seeing.

Meanwhile, the builder forms ideas and pictures from your description according to their own personal point of reference. Maybe you want a kitchen island and they've just completed a project that included an island; the likelihood is it will influence their thinking. Or maybe they always prefer a peninsula so that image pops up instead without them even realising.

The big question is: does your picture in your mind match the picture in the builder's mind? Maybe it does and maybe they bear no resemblance to each other. There is no way to know.

So your job as customer/client is to be really clear what you want and work out how to get your ideas across to someone you've never met and have no idea how they think. And then to keep checking, keep demonstrating and showing any pictures you can find.

It's a challenge, but one that will focus your mind at just the right moment in a build.

Expect the unexpected

And of course – however good the planning – the unexpected is going to occur. So prepare yourself.

It's not possible to start pulling apart an established building without encountering some surprises. No one can know what the state of the wall is under that plaster or how robust the plumbing system is until they eyeball it. The truth will out as work progresses and any unforeseen outcomes just have to be dealt with.

Your best option is to agree how to manage extras before you start, saving the pain of being landed with unexpected bills. Extra work will come up, especially in a retrofit, which uncovers all sorts, so being prepared makes life a lot easier. Communication is key in preventing any dispute. Having a contingency sum set aside will help ease the pain of any hidden extras. (See chapter 9)

How builders work

Builders work in different ways, in part because there are few professional standards to set the tone. Many will have started from a temporary job they liked, learning as they went along. Some will be following in family footsteps and others will have undergone formal training.

How they charge and what they charge will differ according to their style, way of working and relationships with other providers. And it's for you to work out which style suits you best.

It's important to remember that when you first speak to a builder or building professional, they'll be in selling mode. You're a potential customer, so pleasing you will be high on the agenda – they want you to buy from them. You need to look beyond that first charm offensive for clues about their day-to-day style of working.

The costs of any building work are high, so some providers will cut their estimate to the bone in order to get your work. When comparing quotes, think about the structure of the businesses you're considering. A larger company will have bigger overheads, so will factor in higher margins, all of which will get rolled into your estimate. Others will be small and need less profit.

There are pros and cons in both cases – see more below – so a rounded assessment of the estimates and quotes you get will always be the most helpful.

VAT - here's an important one. Certain retrofit works such as insulation, solar panels or heat pumps, can qualify for 0% VAT when the installer both supplies and fits the materials. It's worth setting aside time to check which parts of your project qualify, as correctly applying VAT can take up to 20% off those costs.

HOW BUILDERS RUN THEIR BUSINESSES

There are a number of different working models when employing builders:

- **Building firms:** Small to medium building companies will take on the work of a home retrofit or renovation, picking up once you have your planning permission sorted out.
- **Individual tradespeople:** You may have friends who know a great builder you'd like to work with.
- **Design and build:** Design and build companies deliver every element of the job from start to finish.
- **DIY:** You do all or some of the work yourself.

- **Project management:** You take on managing the project.

BUILDING FIRMS

If you are someone who likes to be hands-on, a small building firm may work well for you.

A small building firm will usually be made up of the business owner and a few employees to cover the relevant trades. They don't get involved in the early stages of the project, just come in when the actual build needs to start.

The most common entry point is when the architect has drawn up plans and planning permission has been granted. At that point building regulation drawings are available to provide a basis for the quotation.

You can draw up your own plans or use an architect or designer to create a design, then work with the planning department to get any permissions you need. It will be for you to decide how involved you want the architect to be. They can just hand over the plans or continue to work alongside as the principal designer (see chapter 8) taking overall responsibility for the project. At a cost, of course

Part of this stage is to work out whether a structural engineer is needed. If you're making any major changes to your house – taking out walls, moving doors, going up into the loft – you will need a structural engineer. They draw up detailed plans for the builder to work to, ensuring your house doesn't come to grief.

Once all that's done, then you look for the building firm who can work to the agreed plans and designs.

At the outset your dealings will be with the business owner. They will:

- discuss the project with you and review the plans you have
- provide you with a detailed quote for the work to be done
- agree timelines
- determine who will be project managing – it could be you if you have the time and the appetite, and if the builder is willing. If you fancy doing this, please be realistic. If you can't manage it, you could cost yourself in time and money.

Some builders will want to do their own project managing and they will have included it in their quote. Others won't offer this service, preferring to bring in someone else to do it. Alternatively, you can appoint a separate project manager. Just remember to factor in the extra cost.

Once all the agreements have been made and the work begins, you probably won't have much contact with the owner. Clarify this at the start so you know what to expect and how to make contact if you need them for specific decisions.

The actual work will be done by employees, with the owner popping in periodically to check on progress and manage any challenges. You will, of course, sort payment with the owner.

This arrangement leaves space for you to be more involved if you want to. You'll have managed the architect yourself. You may have overseen the process for planning permission. All of which will make it easier to talk with the builder, because you're familiar with the detail.

Who will be on site?

Employing tradespeople gives the business owner the flexibility needed to get the job done. But they also don't want employees just sitting around. They need to earn their keep. They achieve this by running more than one job at a time – it means the trades can be shifted to where they're most needed without any downtime.

It makes perfect sense for the builder, but may not be so great for you. If another project becomes time-critical, it'll be all hands on deck there, leaving you with an empty house. I've experienced this numerous times on other projects. On the whole, I didn't mind – and might even have appreciated a quiet day – as long as I knew to expect it. And as long as I knew when the work would begin again.

Preparing for your first conversation

Before speaking with any builders, be clear about your vision – what you want your home to look like, feel like and sound like – and have found a way to describe it.

Then focus on any questions you want to ask the builder:

- Are you familiar with reading architects' plans? If so, what might you need from the architect as the work progresses?
- How many other jobs will you have on the go and how will this affect the length of time it takes to complete our job? How will you inform us of down days/when work will begin again?
- How much cost have you factored in for contingencies?
- Which items are prime cost (definite costs) and which are provisional sums (possible costs)? This should be made clear on the quote but, if not, check it out.

- What issues might you expect that we haven't thought of?
- Who will actually be working in the house and – if not yourself – how often will we see you to answer questions and sort out problems?

You'll probably think of other questions so as soon as they come to mind, write them down. Don't ever underestimate the brain load of retrofit and renovation!

Always get a few estimates to see how they compare. This will also help you clarify the most significant questions, because everyone will respond in a different way according to what they see as important or where they've had problems before.

Managing the unexpected

I hate those moments when a new problem comes to light. It's inevitable – no one can be totally sure what they're going to find when working on an existing building. And there is little to be done other than get on with sorting it all out.

The horrible bit is money – isn't it always!

- *Who is responsible for the additional work?*
- *How much is it going to cost?*
- *When can it be done and how long will it take?*

Given work is already under way there's not much chance to negotiate – you really are in the hands of the builder.

We had this on a number of issues in our own retrofit. And we were not very good at keeping a record to refer to. So please learn from my mistake – keep a notebook with details of every new demand so you can speak confidently about what actually happened.

We had to manage this with our build. A lot of new and unexpected work arose so we knew we'd got more to pay. But then the bill was much bigger than we expected and without a lot of detail. Which left us with two options: pay up and feel resentful or talk it through and look for a compromise. Given there was a lot more work to go, we really didn't want resentment to enter the equation, so we took on the tough conversation.

I remember it well – sitting on boxes and upturned buckets in the building site kitchen over coffee and a cake. It wasn't pleasant. We all aired our thoughts, accepted where we'd both messed up with the money and, thankfully, found our way through to a mutually satisfactory solution. Phew!

SOLE BUILDER

You may prefer to work with an individual builder. This is tempting when you've heard good things about someone who fits well with what you want to do.

This is likely to be a builder who has their own area of expertise and calls in other trades when needed. It's a loose community of people who work together, sometimes as the lead builder, sometimes as a support.

The advantage of this style is that the price can be quite a lot lower than those who have employees that need to be kept working. If small enough, a sole builder may not be VAT registered which will save you 20%. But check at the outset - 20% is quite a lot of extra money to find when the bills come in.

The downside of working with a sole builder is that you're relying on someone to be great at their particular trade and also a good project manager. And if any of that fails, you can find yourself picking up some organising yourself. Of course if you enjoy doing that, then you could take it on from the outset.

Preparing for your first conversation

All the same questions apply as well as providing quality information and plans. In addition it's worth asking:

- What contacts do you have with other trades – plumbers, brickies, electricians, labourers etc. – and how easy are they to access?
- How used are you to reading architects' plans and working with architects?
- Do you have easy access to the building inspector – has this ever been a problem for you?
- What happens if you are unable to work – do you have anyone who can pick up for you?
- What insurance do you have?

If you are someone who likes to be involved, working with a sole builder is a great solution. You're not doing the actual work, but you're able to keep an eye on day-to-day progress. Of course you'll need to build a strong relationship with the builder, ensuring they are also happy for you to be involved and included in decisions.

DESIGN AND BUILD

The clue is in the title. These companies start working with you from first thoughts. They'll have an internal designer who can help work out what you want and produce plans.

The major benefit for you as the homeowner is that you just have one entity to work with. This may be coordinated by an architect or a builder. They'll take full responsibility for every step of the process from working out designs right through to you taking delivery of your newly renovated home.

With this set-up, you can pretty much sit back from the process and just focus on being a good customer – making decisions in a timely fashion so they can get in the taps, bath, kitchen you want to suit the time frame. You won't need to concern yourself with the different trades, drying times or deliveries – all that will be done for you.

Preparing for your first conversation

You want to understand how the company works, what your responsibilities will be and who you'll have most contact with. So ask:

- Who will listen to our requirements and help us work out the design we want?
- Will you produce structural plans, work out the drainage and foundations etc.?
- Do we have to be involved with planning permission?
- Who is responsible for planning out the process of the work?
- Are you able to bring in extra trades if needed – could this hold up the process?
- Who will be our regular contact and how often will we meet for updates on progress?
- How long will you stay involved to manage any snagging problems?

For this to work well, the builder needs to see your vision as clearly as you do yourself. They will price up the job as closely as they can to maximise economies of scale: ordering materials in bulk, making best use of their employed tradespeople when on site, ordering the whole project correctly.

Benefits of working with design and build

The benefit of working with a design and build company is that they should be used to factoring in contingency – extra money in case something goes wrong. They'll have faced most problems before and have a pretty good idea what to expect. In taking on the whole project they know the buck stops with them and they can't keep coming back for more money. It should mean that you can rely on the quote they give you.

The downside of working with a design and build company is that their quote might scare you to death just because they'll have added extra for any problems that might occur.

The trick is to hold your nerve. And ask plenty of questions:

- What can I expect to pay for in addition to your quoted sum?
- Do you have a time scale and what happens if the work goes over? (There are some contracts that penalise the builder for delays but this depends on the wording.)
- What unexpected events do you think likely on this job? Have you added a contingency cost for those? If so, what happens if it's not needed?

If you're able to speak with previous customers (always a good idea) ask: 'who paid when something unexpected came to light?' The good companies will have made an informed guess to cover the unexpected, so they can stick to their quote.

Of course, this won't include new ideas you have along the way. You have to be ready to pay for those. Just remember to ask how much before you hit go.

GETTING YOUR HANDS DIRTY – DIY

If you're someone who loves to get in among it and get your hands dirty then a renovation/retrofit will be the perfect for you.

When it comes to cost, in theory doing the work yourself should make the process a great deal cheaper. However, it's still worth giving some thought to the actual cost.

- **What's your time worth?** Most people forget this one or don't want to think about it. The easiest measure is your value at work. If you work on your home full time then it will cost you in earnings. Is it a sensible trade off for you to stop work in order to save money on a builder?
- **How long will it take you?** As an amateur will it be cost-efficient for you to learn on the job? What will take you a week to work out might take only one day for a professional.
- **How good are you at multitasking?** Are you efficient and able to manage a wide range of issues all landing at one time? Will you manage ordering materials and working out an timed plan for all the different tasks, making homeowner decisions about plugs, flooring and insulation alongside learning how to mix limecrete?

One element at a time could be simple enough for you to take on. The question is whether you're up for taking it all on over a long time. So this needs some honest self-reflection and a willingness to be constantly learning on the job.

Of course if this is something you've always wanted to do, then go for it. It will certainly be very satisfying to sit in your warm, comfortable home knowing it's all down to your own effort.

Just remember, you always have the option to do some of the work yourself while getting professionals in when a task becomes time-critical. What's very exciting in its newness can also wear extremely thin as you approach a second winter with no heating.

PPOJECT MANAGEMENT

If full DIY is too much but you still want to be involved, then you could decide to project manage. This will involve:

- getting in different trades at the right time
- making sure you have all the materials on site when needed
- managing compliance, building inspector, standards, payment etc.

If you're already in the trade or you're an informed enthusiast, this could be a great way to be in at the sharp end without needing the skills to actually do the work.

If you're considering this, do pause and give it serious thought. It's not an easy job. There will be plenty of stress as deadlines loom and you want to get into the house. You have to be on top of materials, protective clothing (PPE), coordinating trades etc. So it's not for the fainthearted.

But if that's the way your mind works, you could get a great deal of satisfaction and save some money by becoming project manager.

SCOPE OF WORKS

Whatever type of builder you work with, a scope of work and contract is a good way to protect yourself. This is a detailed contract that outlines tasks, responsibilities, deliverables, timelines and standards for the project (see chapter 9).

Who manages building control?

The building inspector has to visit the site at key points. Their sign-off is needed for work to progress to the next stage. Whoever you decide to work with, make sure you know which person in the team will engage building control.

Be clear that you want to be kept informed – to know when they have visited – so you can track progress and know exactly what's happening next. It has been known for builders to say they've had the inspectors sign off when they never actually got round to it. And that just leaves a big mess for you to sort out

HOW DO I PAY?

Builders aren't always clear about this part of the process. It's not that easy to talk money so messages can get mixed up and misunderstandings occur really easily. So this is definitely one to clarify at the outset. Not only so the builder is kept happy, but so you know when to access money and you're not caught unawares.

Timely payment is extremely important for a couple of reasons:

1. workers have to be paid on time
2. materials have to be purchased in advance

The larger the company, the longer the gaps will be between payments, purely because they're less likely to have cash flow issues.

Building firms: payment times should be included in the contract with a building firm. If you're not clear how much you'll be asked to pay and when, request a schedule to be added before you sign on the bottom line. As mentioned previously, money is the biggest hiccup – it will always cause problems if you don't comply with what is needed and you can only do that when you have suitable warning.

Sole builders: this will be more ad hoc and may be more frequent, but just as necessary to pay as soon as you're asked. Materials have to be purchased up front, especially if you're using natural materials. The builder may ask you to pay the provider directly, or pay them at the time so they can use their account with a local merchant. It is also possible that a sole builder will ask for cash. It's up to you whether you want to pay that way, but do sort it out up front.

Design and build: the payment schedule will be included in the contract so you'll have plenty of warning. You'll probably be asked for a smaller number of large payments – may be once a month or just two or three times during the build.

PAYMENT QUESTIONS

There are just a few additional questions that it's good to ask, whoever you're working with, because they can cause problems:

Who pays for the skip?

It took me a while to understand the implications of this question. Skips are expensive and if you're doing a chunky build you'll need a few. If the builder is paying for the skip, then they're within their rights to add rubble and rubbish from other builds. If you are paying for the skip, that's the last thing you want – the sooner it fills up, the more skips you need. No reason why you should subsidise another homeowner or the builder.

What about the cost of scaffolding?

A big cost in any build is the scaffolding. The main expense is in putting it up. If it hangs around once it's no longer needed, don't worry – you won't be paying more. It's up to the scaffolder to come and get it.

Just check that the builder has worked out the schedule elegantly so that you only pay once. What you don't want is to put up scaffolding again at a later stage because something was missed. So check the work programming to make sure the structure is being used efficiently.

Who pays the trades?

This question will only be relevant for some building firms and sole builders.

I am often surprised at the need for immediate payment. I was used to having a bit of time to pay bills, so I expected the same. But it's not.

Payment to trades will be expected at the time or at least at the end of the week. So find out in advance. Are you paying the different people or do you pay the builder and they sort it out?

It may be that paying the separate trades directly saves a bit of money – the builder may add a little on top if they're expected to do the admin. Either way, I'd recommend you keep a record of money going out for your own peace of mind and in case there are any disagreements at a later stage.

When do you need payment?

This is a pivotal question in any builder–customer relationship. In my experience, be ready with the money as soon as or before you're asked, otherwise it could be your downfall.

We asked the question of the builder who did a previous renovation. We knew he wanted cash and we needed to get it out of the bank.

'No problem,' he said, 'when you're ready'.

Don't know about you but when life is very busy and you're juggling numerous balls at one time while living in a building site, it can be easy to lose track. Which is just what we did.

The builder didn't ask again, just silently fumed – and sadly we didn't notice. He then told our daughter that he'd never work with us again because we didn't pay up.

Sounds terrible. But we were talking a week, which is long in the building work, but still not terminal. Eventually we realised something was up, asked about it and got a curt reply that put us on the right track.

Ever since then I've made sure money is available. And I've never trusted anyone who says 'it's OK – when you're ready'. That sends me straight to the cash machine!

Withholding money

What about if work isn't done to your satisfaction? Is there ever a case for withholding money?

This is a very difficult one and it comes back to the relationship you have with your builder. Set up a norm in that first interview by asking how they manage disagreements and problems. This is important because you might need to:

- ask for something to be redone or changed
- question why decisions are being made in a particular way
- point out they've not used your chosen materials and have them swapped.

The list is endless so there has to be a means of resolving conflict.

It's not unusual to set a time clause into a contract where money is retained or even taken off completely if the timeline isn't reached. If you have specifics requirements you could treat them in a similar way – make sure they're included in the contract with clear penalties. If you get the relationship right you'll never need them but better to be safe than sorry.

The final invoice

Final payment for the work will hang on the decision about when the work is completed.

Will you wait for the completion certificate or is it the day the builders leave site? These may not happen at the same time. And who will make that decision?

And what about snagging – all those irritating little glitches that need repairing, replacing or redoing? Do you have to pay for this? Be really clear about this at the outset. It's the sort of problem that can damage a relationship regardless of how well it's gone until that point.

FINAL WORD ON COST

You'll have a budget and an estimate of cost from your builder. You'll have worked out where the money sits and how you can access it when you need it. Great planning.

And it's unlikely to be enough.

Put aside what you can as a contingency because the chances of staying on time and on budget are minimal. I've heard of very few that went exactly as planned.

As a rule of thumb I add at least 25% more in cost and probably similar in time. Depressing, but realistic. Manage your expectations and don't give yourself the stress of an overrun.

Summary

- There are different building businesses, from building firms that pick up once planning permission is obtained or plans signed off, to single builders who do the work bringing in other trades when needed, and design and build companies that do the work from start to finish.
- Be well prepared, with a very clear picture of what you want, before asking for estimates and quotes. It saves time and gives you a better idea of what the total cost might be.
- Builders will ask for payment in different ways. Whatever you agree, always be ready – they will have bills to pay on your behalf.
- Get clear on some specifics – who pays for the skip and the scaffolding. Has it been included in the estimate?

- Whatever work you're doing, it will always cost more and take longer than planned. Manage your expectations and put some contingency money aside.

To find the information indicated by the * follow the QR code.

NOTES

CHAPTER 8
THE PROFESSIONALS THAT CAN HELP YOU

Few of us ever do a renovation or retrofit project on our own. As soon as we get the idea we start thinking about who can help. The first port of call is often the builder, but others will have to be involved if the work is to go smoothly.

I can almost hear the cry: it's going to be expensive enough already. Why would I pay out more when the builder can do it or I could do it myself? Of course, that's correct at one level, but let's take a look at the help that's out there – both the people you actually need and those you might decide you want.

The scale of the work will determine who needs to be involved. The larger the renovation, the more people you'll need. A small-scale piece of work might just be you and the builder.

Who do I need?

I've discovered that knowing who to approach for professional help when planning a build is a minefield because of overlapping roles and overlapping duties. And whether anyone will tell you about the different options is questionable. It's back to being in sales mode – they may not want to suggest others who could be better suited.

There'll be some people you have to get on board and some who'll be a huge help if you can afford it. The relevant people will vary slightly according to the size of the task before you.

How big is big?

How big is the job you want done?

For most of us whatever we do will feel 'big'. Even a very small change can feel huge if you don't like disruption, while the person used to major renovation work will think nothing of adding in a new kitchen. So it's partly a matter of perspective.

Renovation is classed as large-scale when you include:

- knocking down walls
- adding in steel girders to hold parts of the house up
- moving staircases
- going up into the loft
- adding new rooms.

When you get to this level of change, you'll need professional help. Knocking down walls requires the skills and knowledge of a structural engineer; adding extensions is best worked out by an architect or architectural designer; major changes to the house will automatically fall under building regulations, in which case you need either a builder who's capable of managing this for you or an architect.

It's a matter of knowing who does what and who you need when, so you can make best use of the experts without breaking the bank.

THE PROFESSIONALS THAT CAN HELP YOU

Principal designer (PD)

Before we go into the question of building professionals let's get this one sorted out. This is a relatively new role, designated by government, that has oversight of any construction project. The person who takes on this role has legal responsibility for health and safety issues on site right through to the safe delivery of the project according to building regulations*.

So far, reasonably clear. But who is the PD? In practice, the PD role is often taken on by an architect, architectural designer or sometimes a builder with the right competence and insurance.

It's even possible to do it yourself – if you like detail and admin, and you have the time. There's some paperwork, correspondence and emails to deal with and you'll need to get up close and personal to the project. If you want to do that, go explore health and safety and make sure you understand building regulations.

In truth, you will always have ultimate responsibility for your own home. The buck stops with you – as it should. It's your home and you know how you want it to work. So even when you have an official PD – someone you have checked out thoroughly to make sure they are competent – you still need to be actively involved.

WHO ARE THE BUILDING PROFESSIONALS?

Who does what and when do you need them? It's not as straightforward as you might think and there's a difference of opinion even within the building profession. I'll do my best to cover the different roles and what you can expect from them. Hopefully that will give you the information you need to ask the right questions so you can get started.

The relevant people are:

- architect
- architectural designer
- interior designer
- structural engineer
- building inspector
- retrofit coordinator.

Architect

'Let's build an extension – what do you think?'

How many of us have had those conversations! It's a quick question that opens up a whole can of worms. First hurdle being: 'do we need an architect?' The second thought is usually: 'they cost a fortune!'

It's back to the size of the job. If your plan is to gut the house, knock down walls, add full insulation… then it would be prudent to employ an architect. It's not essential – and builders may say you don't need it – but it can stop costly mistakes at a later stage. They can also take on being your principle designer if you want them to.

Then there is the question of planning permission: whether you need it or if the work you want done comes under permitted development. If you've had an architect design your extension or loft they'll know what you need to do.

If you do need planning permission, they will:

- create a design that provides what you want and will pass local regulations
- check in with the planning officer during the consultation period to find out about possible objections

- adjust the design accordingly.

If they do all this then submit the plans, there's every chance it will go through smoothly.

Of course you can do it yourself – but make sure you enjoy admin. If not, and you can afford it, getting help will be easier, quicker and more likely to succeed.

The key difference between an architect and an architectural designer is registration. Only those registered with the Architects Registration Board (ARB)* can legally call themselves an architect. Becoming ARB-registered usually takes around seven years, including undergraduate and postgraduate study, supervised practice and a professional exam. Architects must also complete continuing professional development (CPD) to maintain their registration.

ARB-registered architects must carry professional indemnity insurance* to cover negligence, errors or omissions.

There's more detail about the planning process in Beginner's Guide to Eco Renovation.*

Architectural designer

Architectural designers are not registered with the ARB. They may have studied architecture, architectural technology or a related field, or gained experience through practice.

Architectural designers may also carry professional indemnity insurance, but it's not a legal requirement.

(**NB:** While we're on the subject – it is always wise to check professional indemnity cover with any professional you hire.)

So why would you choose an architectural designer (AD) over an architect? The biggest reason is probably cost. An AD is well equipped to take on a residential project – they may even have more experience of what you want – and they'll do it at a much lower cost.

Interior designer

This is where it starts to get complicated. My understanding was that interior designers came in once all the mucky work was done to help you make your place beautiful.

In conversation with an interior designer recently, I asked, "Who does the George Clarke bit of moving the stairs or shifting doors around? Is that an architect?" Her answer – "It's me." Turns out the interior designer will design the inside of the shell and work with the builders to achieve it.

SO much more than I expected. Interior designers must get so many confused requests, because everyone I've asked so far has had the same understanding as me!

Could we afford an interior designer?

It was a question we asked ourselves for a while when doing the last but one renovation on our home. We were finally stripping out all the battered old furniture that had raised two kids (plus three dogs, one cat, two rabbits and a guinea pig) – it was time for an elegant home that suited our status as empty nesters.

I assumed I was looking for an interior designer. In fact what I wanted was an interior decorator – someone to help with the look of furnishings, not where the walls would be.

I went searching (using all the wrong terms) – just out of interest – and managed to find someone who could help us plan out the look, make suggestions, source the goods – even take us shopping. Then she'd buy the goods, taking the difference between her trade price and the amount we'd have paid anyway. Cost to us? Zero.

Seemed like a good bet, so we went for it. It worked really well! Lianne helped stiffen our backbones. She cajoled, encouraged, sometimes goaded – made us think about what we really wanted. Until then, I'd been very safe in my choices. With her help, we now have the lovely comfortable home we want. We're both more adventurous – it's only once gone wrong and that was reversible.

Now I've finally understood. An interior designer can do far more than select finishes and furniture: they plan layouts, advise on flow and design the inside of the shell. They can prepare drawings for non-structural changes, but where load-bearing walls, staircases or other structural elements are involved, they will still have to involve a structural engineer and sometimes an architect.

A good interior designer will be familiar with relevant regulations and, where needed, can prepare drawings for builders or work alongside other professionals to ensure compliance.

The biggest difference will be in cost. You're not paying for the long years of training put in by architects so, if your work isn't structural, you could get what you need for less outlay.

It's your design

Quick point before we move on: this is your home and your design. So don't be polite or accept something that isn't what you want. All of the professionals above can give you a design but you have to live with it.

When they give you the first pass at plans, take your time to review them. Imagine how this design would work, mark it out on the floor, draw your own sketches of how it would look.

Think about where furniture might go – anything to help you get a sense of the final look.

Some people can do this easily. For others of us, it's a challenge. I really struggle to imagine what something might look like and it's tempting to assume the professional knows best. But they are not you and they may have different tastes or priorities. So take your time and make sure it's what you want. You have every right to go back and ask for it to be changed. As often as you need.

Structural engineer

Any change that affects the structure of your home will need a structural engineer. This is the person who does the calculations and plans that ensure your house won't fall down. Pretty significant, then!

It applies to anything affecting structure – taking out supporting walls, putting in steels, removing doorways, opening up the loft – anything that could damage the integrity of your house. The structural engineer provides calculations and drawings, which must then be submitted to building control for approval.

Your architect, designer or builder will confirm if a structural engineer is needed and they'll probably have someone they're used to working with, which can make the process a lot easier. They will provide structural details and specifications for the builder, showing how supports, beams or alterations must be installed. These plans will be shared with the architect/designer and with the building inspector as required.

Building inspector

Which brings us neatly to the building inspector. Unless you're closely involved with the build, you'll probably never know about these visits. The builder may have a relationship with the local authority building inspector, which is helpful because it can make the process smoother.

The building inspector is required to inspect each stage of the build and keep a record to show the work is compliant – when foundations are completed, when the steels are put in, when the roof goes on... They check that the work has been done according to the regulations and the plans.

This is obviously important for the wellbeing of your home and it's required before the build can go to the next stage. This is where the relationship between builder and inspector is useful. Our builder knew that he could call and the inspector would come either that day or the next so there was no hold-up.

Private inspectors* can be a good fit for retrofit projects, as they may take a more flexible and solutions-focused approach than local authority teams, particularly when dealing with older buildings and bespoke details. If they'll be acting as the registered building control approver (RBCA), make sure they are properly registered with the Building Safety Regulator.

Once appointed, either you or your builder will need to submit an Initial Notice to your local authority to let them know you've chosen a private inspector. This notice must be signed by you because, as the homeowner, you carry the legal responsibility for ensuring the works comply with building regulations, even if your builder or designer is managing the process day to day.

Project manager

Someone has to take on the role of project manager. This is the person who ensures materials are on site when they're needed, who gets the trades in the diary in the right order so the work isn't held up, and the person who manages payment.

If you are working with a larger firm, a site or project manager may be part of the package. In a small domestic project, this role could default to you, unless you've specifically asked for project management to be included. Builders will manage their own trades, but not always the client-side tasks e.g. ordering kitchens, window fitting, managing other professionals etc. If this does fall to you, take it seriously. Not having the materials you need at the right time or having the kitchen fitter arrive before the floor's finished will cost you in time and money.

If you don't want to take this on yourself, talk to your professionals. You may find this is something they offer – just remember to add it to the cost.

Retrofit coordinator

If you're setting out to retrofit or you're including it alongside other work (and of course you are!) a retrofit coordinator* can be a real help. The clue is in the title – they'll help you identify ways of making your home energy efficient. Their role is regulatory and compliance-based under PAS 2035 (see chapter 2), which means they can provide you with a plan of action that fits all the requirements of PAS 2035 – the British standard for domestic retrofit – including risk assessment and sequencing.

There are a couple of options with this role:

- **Retrofit assessor:** this person will survey your home, according to the framework for PAS 2035, (see chapter 2) gathering data about day to day functioning. This will be done in a lot more depth that a standard survey. Once they have the data they need this is passed to a retrofit coordinator.

- **Retrofit coordinator:** using the data gathered by the assessor, the coordinator will produce an assessment report that tells you what you can do to improve the energy efficiency of your home. They will cover everything you need, except heat calculations for a heat pump and data for a ventilation system.

If you are keen to improve the performance of your home, but don't want to be hands-on, talk to the retrofit coordinator to see if they can offer more help. Depending on their approach, some are happy to take on day-to-day project management.

There are also retrofit companies that specialise in start-to-finish support, from updating your EPC, retrofit assessment and report, finding trusted providers and project management, right through to the end result*. This is particularly helpful if you have builders who aren't clued in to energy efficiency.

If you want to go down this route, start by looking for a retrofit coordinator. They may also be a qualified assessor. If not, they can bring in an assessor to gather the data for them.

Warning: whatever they claim, not all retrofit coordinators will be environmentally friendly. True, they can help you be energy efficient now, but they may not help you to future-proof. So read chapter 11 before you speak to them, then you can ask the right questions about materials.

Expertible: if you are stuck for which way to turn, take a look at Expertible. This is a community of experts who will have a 1-1 call with you at a small cost and help get you on the right track. Follow the QR code for the link

SAFEGUARDS

Much as you love your house right now you might decide to move one day, so make sure any renovation work will facilitate that. As we all know, a big part of the selling process is the searches and part of the searches will be to check up on any building work or structural change you've made during your tenure.

Planning permission: if you needed planning permission for the work you did, then this will detail what was done and how.

Certificate of lawful development*: this is worth having when you don't need planning permission. It's a legal document from your local planning authority, confirming that the changes you've made came under permitted development. It's useful if you're concerned about neighbours or for when you want to sell. It confirms the work was lawful at the time and protects you from later enforcement action. It also clarifies to prospective buyers and mortgage lenders that everything you've done is above board.

Building Control Completion Certificates*: these are documents from the local authority or approved inspector confirming that construction work complies with building regulations and has been inspected and approved by a qualified building control surveyor. They are crucial when selling a property because they demonstrate that the work was inspected and approved for compliance with building regulations. Although they're not a guarantee of workmanship, most buyers and lenders will require them.

Natural materials

Unless you've found specific eco professionals, you will need to stipulate the materials you want. It's not only builders who are in the dark about environmentally friendly options such as insulation materials, low-VOC paints or ventilation systems. Architects and designers can also be out of touch.

Insulation must meet the performance standards in the Building Regulations 2010 (Part L)*, though you can usually choose the specific material you want as long as it achieves the required U value (see chapter 6).

Remember that builders are keen to get the job finished, so they can bill and move onto the next one. This can make them risk-averse when it comes to stepping outside the box. Getting your architect or designer on board can set the whole project up for success by:

- specifying specific materials in the plans
- talking to the builder about what's needed and answering any concerns they have
- visiting to make sure the work is being done as prescribed.

And of course, make sure you're fully informed and involved. One major learning for me writing this book has been the importance of natural materials in future-proofing your home (see chapter 11) so you can be warm in winter and cool in summer. So do your research, speak with experts and be ready for objections. Remember to emphasise the benefits to the builders for their own health and wellbeing. No need for protective gear with natural materials!

*When they try them, they like them**

Recently I visited a home retrofit site to chat with the builders about their experience of working with hemp insulation, blown glass aggregate, hempcrete and wood fibre etc. It was an interesting discussion.

They're a well-established small building firm used to working on extensions and major residential alterations. As you'd expect, they normally worked with PIR insulation, plasterboard and gypsum plaster. But this time the homeowner was a retrofit coordinator totally committed to eco builds. So they took on the challenge to work with natural materials.

By the time I met them, the bulk of the work was done and the clay plastering (done by specialists) was starting the next day.

The owner of the company told me that they'd been unsure at first. It had taken more organisation as they got used to a different way of working. But now they had the experience they were keen to work with the materials again. 'It's much better for us to work with – none of the nasty chemicals we normally have to deal with on a job.'

So when they try them, they like them. And the more we can encourage builders to experiment, the more informed they'll be and hopefully they'll realise how amazing these products are.

THE PROBLEM WITH BUILDING REGULATIONS

This is important to know. There is a problem with building regulations that gives an easy out for builders who don't want to change.

Regulations require a specific U value for insulation used in a renovation/retrofit. There is no problem with natural insulation materials being used as long as they fit the required U value. So far so good.

But the industry defaults to PIR because it is efficient without being thick and it's easy to get hold of. According to building regulations, there's no problem using natural insulation as long as it matches the required U value, condensation risk, fire performance etc. It's all possible, but for a builder not familiar with how natural materials are measured, the increased thickness needed to reach the required U value can be a blocker. In a loft extension, for example, this might bring the ceiling level down too low, so they won't use it.

But the regulations are missing out on two vital pieces of information:

- When measuring natural materials, it's the whole system that matters. So the specified thickness of hemp when coupled with the other factors of retrofit – like airtightness – will deliver the required U value.

- U value measures how well the material keeps heat in during winter. Low U value won't keep heat out during summer. That requires decrement delay and thermal mass, neither of which are included in the regulations.

So building regulations don't incentivise the use of natural materials, but they don't ban them either. It's for us homeowners to stick to what we want and bring the builders with us.

Give building regs its due, details are updated constantly to keep pace. The downside is this requires builders to stay on top of each change and, since they are so busy, they're more likely to stick with what they know.

You can work around this with a forward-thinking designer who understands that just complying with legal measures may not be enough. Agree with them on the best materials for future-proofing your home. Then, once the plans are drawn up, they can check with building regulations to make sure it is all acceptable before it goes to the builder. With the right expertise on your team, you'll be able to get what you want.

> *Passive House training helps*
>
> *We are a perfect example of where natural materials have worked and been accepted by building regulations.*
>
> *We employed an architect firm that included a partner who was Passive House-trained so we could be sure of using planet-friendly materials and being energy efficient. As a result, the plans for our extension included: wood fibre insulation set within a wooden framework with an airtightness membrane behind and a compacted wood fibre 'plasterboard' on top.*
>
> *If building regulations took values for the wood fibre insulation in isolation, it wouldn't pass. But together the system delivered the required U values.*

So if you're told you can't have the natural material you want, don't just accept it. Find out more and start asking questions.

It's not unheard of for builders to agree to the use of hemp or wood fibre at the estimate phase. Then, once you are tied in and the work has begun, tell you that building regulations won't accept the insulation you want, so sadly you'll have to have PIR after all. So it definitely pays to do your homework and find the relevant professionals to refer to.

Summary

- The principal designer takes overall responsibility for the work. You always share that role because it's your house and you are responsible for what happens.
- Architects, Architectural designers and Interior designers can all help work out changes to your home. Professional training will vary, and so will cost.

THE PROFESSIONALS THAT CAN HELP YOU

- Major changes will always need a structural engineer and building inspector.
- It's worth understanding building regulations, especially if you want to use natural materials
- Whoever you work with, be clear if you want to use natural materials – don't assume they will include this.
- If you get stuck, contact Expertible for a 1:1 call with an expert. (Follow QR code)
- If you're a woman retrofitting and you'd like an informed support group, try Her Own Space*.

WHAT THE BUILDER WON'T TELL YOU

To find the information indicated by the * follow the QR code.

NOTES

CHAPTER 9
WHAT YOU NEED TO KNOW ABOUT CONTRACTS

We're used to making contracts in life – we sign them when buying a car, a house or sorting out insurance. But it's remarkably easy to forget when working with a builder. Yet the implications are massive. The builder could:

- make major mistakes in the build and leave you with a substandard home
- go bankrupt during the work and never actually finish
- not follow the agreed plans so you don't get what you want
- charge large amounts for any unexpected additional work going beyond your budget
- go well beyond the agreed date.

All these and many more events could occur and without a contract you'd be in a real mess.

YOU ALWAYS HAVE A CONTRACT

Let's be clear – as soon as you agree on a price and start date, you have a contract, legally. It won't necessarily be helpful, but it is an agreement that has to be honoured.

The value of a thought-through contract is that it's not based on opinion and memory. A written contract, prepared in advance and agreed on together, makes it much easier to sort out disputes and disagreements.

There are different options for creating a contract – you can:

- accept the contracts provided by the professionals at the beginning of the relationship
- find a contracts expert who will draw up a contract that suits you
- use a template prepared by an expert and complete it with the builder*
- write up each conversation you have with the builder and share it in a joint email
- stick with the verbal agreement and cross your fingers.

Clearly the final option is not to be recommended! This is your home you're talking about – probably the biggest investment you'll ever make – so much better to give yourself some security.

Signalling intent

The great value in producing a contract is that it helps you clarify what you want. You have to get clear in your own mind what you're expecting, what is acceptable and what you don't want. As soon as it is written down and shared with the builder you'll know if you both have the same understanding.

In chapter 5 I talk about the challenge in conveying the picture in your mind. It's so easy to think you've done a great job but discover – often too late – that the other person has a very different picture in their mind. Preparing a contract is one way of avoiding that.

Hopefully you'll never have to enforce the contract, but it'll still be a huge help in exposing disagreements and misunderstandings early on, and is a way to ensure that everyone has the same expectations.

WHAT MAKES UP A CONTRACT

These are the four elements of a contract:

- **Offer:** the builder's proposal, setting out the estimated costs, timelines and scope of work.
- **Acceptance:** your agreement to that offer, including any changes that have been negotiated.
- **Consideration:** the agreed cost or contract sum to be paid for the work.
- **Intention** to create legal relations: the shared understanding that the agreement is legally binding.

The contract offers protection against all the incidentals we don't want to think about, but if you look at it as a document that's part of the planning, it becomes much less of a drag.

What needs to be in a contract

The necessary basic clauses are:

- **Dates:** when the work will begin and an anticipated end date. If you have a date beyond which the work must not go, then this needs to be clarified up front and agreed with the builder. You may also want to consider a penalty clause – a cost for every day over the end date that the work continues.

- **Consideration:** the total sum of the contract and the scope of work to which it applies.

- **Payment plan:** this focuses on how payments will be made. If staged payments are requested they should ideally be tied to completed work milestones, not just arbitrary dates or large up-front deposits. Clarify the style of payment – cheque, direct transfer, cash. (See chapter 7)

- **Termination:** this clause covers how you can terminate work during the contract, should that become necessary. Who gives notice and to whom; how the final amount owing will be agreed and how it should be paid, what will happen to outstanding work, plus any obligation to sort out remaining problems.

- **Dispute resolution:** this is a really good one to work out. The contract outlines what needs to happen if you pass the point of talking it through together. Mediation* is a first port of call – asking the help of an independent professional who can listen to both sides and help you find a way forward. Local laws will apply, so the final stage will be using the courts, but it is always best – and cheaper – to avoid this if you can.

Additional clauses

Depending on the work being done, there are some additional clauses you might want to add in:

- **Waste management:** who is responsible for the management of rubbish and excess materials. Including who pays for the skip and who has the right to fill it (see chapter 7).

- **Change of design:** the intention of every renovation is to stick to the agreed design. However, there are always surprises that need to be managed. This clause explains how the builder will outline what's required, how quotes will be provided and agreement reached.

- **Extras:** this is an important clause. If not agreed in advance it could be the straw that breaks the camel's back. So many good relationships go sour over money, so anything you can do to avoid this is important. You need a clear agreement about how extra work will be charged. Some builders will opt for a daily rate, some an hourly rate. If you agree that in advance, you'll always know what you're signing up for.

- **Liability insurance:** this ensures that your builder has liability insurance against claims of negligence. There is not a statutory minimum, but industry practice is usually cover of at least two million pounds, with two to five million being common. The actual amount will depend on the project.

- **Accidental damage:** accidents are not covered by the builder's insurance so you need to work out how you'll manage them. Tighten a screw too hard and crack a tile, drop a hammer on the new floor and dent it, break a window with a scaffold pole – it's a miracle if something doesn't get broken somewhere. Some accidental damage may fall under your own household insurance, but do notify your insurer before work begins – see below.

- **Agreement about managing accidents:** discussing this ahead of time will make it easier should an accident occur with materials or property. You'll need to be pragmatic but at least you'll have some basis to start with.

- **Retention clauses:** a retention clause is when you withhold a small percentage of the agreed amount until all defects are rectified. This needs to be agreed with the builder at the outset and put in writing. For example, you could include a clause that requires Building Control Completion Certificates* and snagging (sorting out the small bits and pieces that have been missed) before the final payment is made.

Not my fault

Problems happen so easily and often with no direct fault, but it still has to be managed. We had a couple of difficult moments in our retrofit that we should have been better prepared for.

First, when John noticed a chip out of the side of the new bath. It had been plumbed in that day and the only people in there had been the plumbers. When we asked them how it happened, they denied all knowledge of it. If they didn't do it, it's a mystery. But we had to take it on the chin. Fortunately we got a 'magic' expert in who mended it. Never noticed it since.

Second was the day we noticed the insulation around pipes in the kitchen was only on one side. The pipes had been installed too close to the wall, so the insulation tube wouldn't go on. Instead of sorting it out the plumber had cobbled it together hoping we wouldn't notice.

Fat chance! This is why we lived in – to keep an eye on progress and quality. When they came in the next morning, there was an uncomfortable 10-minute discussion before they agreed to take the pipes out and start again.

Not pleasant, but it had to be done. Without it, we'd have been left with a considerable cold bridge that would have affected the efficiency of the room.

Practical help with contracts

It is all pretty daunting, especially when you are putting so much money on the line to work with people you don't know well. It can feel like a minefield.

But there is help at hand so you don't have to rely on the builder's own contract:

- **Joint Contracts Tribunal (JCT)*** has standard contracts you can download. JCT is composed of various industry organisations, producing standard forms of contracts and other important documentation for the construction industry.

- **Federation of Master Builders*** also provides standard contracts to contractor members.
- **The Place Between*** is a social enterprise for home renovators has a template toolkit which is user friendly for both homeowners and contractors.

INSURANCE

Some professionals will be thorough and tell you what you need in terms of insurance. Others won't think about it at all. I've been through four big renovation changes and never been told anything about insurance cover.

Builders and contractors must have their own cover. At a minimum, this means:

- **public liability (PL) insurance:** protects against injury or property damage on site (e.g. a visitor tripping or a neighbour's car being damaged)
- **employer's liability (EL) insurance:** legally required if they employ staff, covering injury or illness to workers.

Professionals: architects, designers and coordinators should carry professional indemnity insurance, which covers accidental damage or injury as well as protecting you if they make mistakes in their advice or design.

Always ask for a copy of your builder's insurance and, if necessary, show it to your insurer so you know every angle is covered.

Builders need their own insurance to cover themselves for negligence. The amount will depend on the project, but will usually be in the region of two million to five million pounds.

Your home insurance: check with your buildings and content insurer and let them know what's going to happen, including whether you'll be living in or moving out. They'll clarify whether your existing insurance will cover you during the process.

Your regular policy is normally good for accidents while decorating or maybe even taking out a chimney, but for anything bigger they'll refer you to a specialist insurer. It's a good idea to get a copy of the builder's insurance and show it to the specialists to make sure you have all bases covered.

And finally

I bet your alarm bells are ringing by now! It's never nice to focus on the negatives. But always worth facing up to the worst so you can be prepared.

Remember you are the ultimate principle designer, so you have choices. Take your place in the team and accept joint responsibility. It's the most productive mindset to have and will make your life a lot easier.

Summary

- As soon as you agree to go ahead you have a contract, but without something written it's much harder to enforce.
- Working on a contract together can help clarify that you are all on the same page.
- It's useful to cover all aspects of the process in the contract and any additional details that are specific to you.
- You can accept the builder's contract or download templates off the web.
- Check out your buildings insurance – you may find it doesn't cover you while the builders are in.

To find the information indicated by the * follow the QR code.

NOTES

CHAPTER 10
WHAT THE BUILDER WON'T KNOW ABOUT RETROFIT

Energy efficiency is an idea we all understand. But use the official word – retrofit – and hardly anyone has heard of that, builders included, so it's time to spread the word.

Retrofit just means adding in what wasn't included in the original build. Whatever label you use, we're just making a building perform better, so you have lower energy bills while being comfortable and healthy.

Hard to believe, but it can be achieved in any home. No matter how cold, damp or mouldy your home, there is the potential for it to be a comfortable, sought-after, delightful home that won't break the bank. So what is actually involved?

THE BIG FIVE

There are five central elements of retrofit:

1. Airtightness
2. Insulation
3. Ventilation

4. Sustainable heating
5. Renewable energy.

Breathability: There is also a sixth element to add to the list for those who live in houses with solid brick walls and suspended floors. This generally means they were built before 1930, as well as some built between 1930 and 1940. If this is your house, then you must factor in breathability.

Breathability, or vapour permeability, definitely isn't on the radar of most builders, so make sure you understand the basics before you start your explorations.

This chapter addresses airtightness, insulation and ventilation, as well as breathability; chapter 12 covers sustainable heating and chapter 13 looks at renewable energy.

Any house can be retrofitted

As long as all the different elements are added correctly, any house can be retrofitted. Of course, success requires builders and homeowners to understand what 'correctly' means. Very few builders are well informed about this, so it's time for homeowners to pick up the slack.

Living in a draughty old home

I've lived in a Victorian home for over 40 years and I can tell you it used to be bl***y cold. Doors and windows were ill fitting, howling gales came up through the floorboards from the sub-floor void (the open space beneath the joists), wind whistled down the chimneys. It was a nightmare and I was permanently wrapped in scarves, multiple jumpers and jackets. Even got a long pair of Ugg boots to wear indoors as slippers.

We moved in during Easter 1979. At the time there was one radiator in the main living room plus a gas fire, a radiator in the 'parlour' (which was never switched on) and one on the landing. When we'd viewed the house it was toasty warm, but we learned later that it was all for our benefit. On a day-to-day basis it was freezing.

As we began improving the house – double-glazed windows, carpets on the suspended floors, additional radiators – we noticed there were a couple of areas where we got mould. Both in corners of the house, one of which faced the prevailing wind. So clearly the wall itself was cold, inviting the warm air to condense.

At the time, we put a second skin on the bedroom wall in the hope that a warm surface would stop the problem. During our recent retrofit that second skin finally came down. We discovered all it had done was hide the mould, which was still alive and well. Warm air had passed through the joins then condensed on the still-cold wall. And the non-vapour-permeable plasterboard we used had stopped any movement of the vapour back into the room.

We had had no idea what to do for the best. Like many before us, we gave it our best shot and hoped for the best. Now I know it was quite the wrong thing – but we learn and do better next time!

Let me explain

There is no problem changing a building as long as we understand what we're doing. So let's get clear:

Older homes: when older homes were built (before 1940) vapour could move passively through the walls and active ventilation was blown in through chimneys, from under floors and around doors and windows. Moisture just wasn't a problem they had to deal with. Cold, yes, but not moisture.

However after WWII, construction changed, because of the need to build houses super quick. New materials were available, courtesy of petroleum – modern, faster-drying materials like concrete, gypsum and cement – that made the job much easier.

It was a perfect time to get a job in construction – the new materials were less complex so the job didn't require specific skills. That opened up opportunities for loads of new workers. The downside is that a lot of traditional skills were lost – they just took too long to learn:

- Gypsum plaster is quick-drying with just one or occasionally two coats versus lime plaster that needed three coats plus drying time
- Plasterboard (drywall) and cement-based plaster were simple to install versus creating lath and plaster ceilings and walls, which is a long-winded, skilled job.
- Prefabricated sections were brought to the site versus building on site from scratch.

The ease of use made it possible to bring on unskilled people and still build at a fast pace. Old skills and old knowledge just weren't valued anymore. So now we have builders without the traditional skills that would enable them to upgrade and future-proof old houses.

End result?

Condensation and mould in old homes that are:

- Draught-free because homeowners have filled, blocked or covered everything they can think of in order to save money or
- improved/renovated' for warmth but done without the basic knowledge of how the building works.

Both options – all done with the very best intentions – leave a legacy of condensation because the focus has been just on warmth and not air movement. Put that together with new materials that don't work in old houses and it's a recipe for disaster.

Modern homes are built very differently, but condensation is still a major issue. Without natural draughts and breathable materials, it becomes essential for modern houses to be fitted with quality ventilation.

If a home is lined with plastic-based insulation like PIR, XPS or polystyrene, you have to find a way to deal with the 20 litres of moisture produced every day by a family. It's got to go somewhere and the plastic nature of the insulation offers no help.

We need energy-efficient homes

Now more than ever we need our homes to be energy efficient and ready for a volatile future. So we have to think about retaining heat in winter and keeping heat out in summer.

It's possible – we can do it. Remember, I'm saving 75% of my previous energy usage in my old Victorian home and I could do even better. So it's all to play for. You just need some information so you can get builders who understand or builders who want to learn.

AIRTIGHTNESS

Airtightness is the missing link when it comes to having an energy-efficient home. When I wrote *Beginner's Guide to Eco Renovation* I had a brilliant retrofit architect as my mentor – Julia Healey*. I remember her telling me: "if you insulate without making airtight you'll be disappointed. You'll still have cold feet in the evening." This is because air has an amazing ability to find each and every gap, however small, and when the weather is cold, insulation on its own just can't save you.

Insulation and airtightness go together like peaches and cream. Insulation keeps you cosy and airtightness keeps out those freezing cold draughts that make life so miserable.

Don't hold your breath waiting for your builder to bring up airtightness – unless you're fortunate enough to find knowledgeable eco builders. For most builders and building firms, airtightness isn't on the agenda, even though it's the piece that will lift you up into true energy efficiency.

You can DIY a certain level of airtightness by cutting out draughts yourself:

- adding draught-proofing tape to your windows and doors
- sealing up gaps between floorboards, blocking up your chimneys, making sure the loft hatch closes tight
- checking the condition of your exterior walls for gaps and breaks in the pointing.

Or you can go for a fully airtight build and add in a vapour-permeable, airtightness membrane inside of the exterior walls of your home before you add insulation and make sure that all uncontrolled air is blocked from entering your home.

Hang on a minute: how can a membrane be airtight <u>and</u> vapour permeable? Turns out water vapour molecules are smaller than air molecules. The membrane has tiny microscopic pores big enough for vapour to pass through, but too small for water or air. Clever!

Cutting draughts makes a big difference to the warmth in your home. Anywhere that cold air sneaks in is going to increase your energy bill. When energy was cheap we could ignore it and whack the heating up, but now sky-high prices mean we want to save wherever we can. Hard to live with, but a bonus for the climate – makes us all more careful.

If you want to reduce your uncontrolled air but can't go the whole hog, then you need to look at where warm air is being lost in your home.

Thermal imaging camera: this is a great way to find out where cold air is coming in. Take a look at your local council* – they sometimes have them to borrow for a minimal sum or just a returnable deposit.

It's best to use a thermal camera when the weather is cold. You need a temperature difference of at least 10°C between inside and outside, otherwise it won't give quality information. The places to check are:

- **Windows:** where the glass meets the frame; where the window closes; where the frame fits into the wall.
- **Doors:** where the door fits into the frame; where the frame fits into the aperture.
- **Floors:** if you have floorboards they'll be sitting over a sub-floor void with airbricks, so prone to draughts. Check the joins of the floorboards and where the floor meets the wall. (**NEVER** cover your airbricks!)
- **Chimneys:** check how much cold air comes down the chimney. If you're keeping it, look for a way to

block it up, making sure you can remove it when you want to light a fire or let in some fresh air.
- **Walls:** 35% of our heat can be lost through the walls. Check the temperature of your walls and look for trouble spots. Check the mortar outside at those points to see if there are any breaks.
- **Loft:** this is generally the first place to put insulation, so you probably have this covered. However, it's worth checking you have the thickness now required by building regulations, that no parts of the loft have been missed; no insulation has slipped or been shifted out of place; your loft hatch fits tightly and has insulation on the back.
- **Inlets:** check around plumbing and electrical cables; air vents that may have a gap between the unit and the wall; any inlets into the home.

For more information about energy hacks, follow the QR code at the end of the chapter and download a free booklet on my website: *Stay Warm for Less**.

Airtight build: if you're already planning a renovation, you have the opportunity to go for full airtightness and create a thermal envelope around your house that blocks any uncontrolled air.

An airtight membrane is wrapped around the external walls of building to make sure that there are no gaps. In essence the membrane seals up the house from the outside space.

It really is the windproof jacket you put on when going out for a walk in winter. We all know that a warm jumper won't do on it's own because the wind just goes through the gaps in the knitting. But put a jacket on top of the jumper and you're cosy. Same principle in building or retrofitting. Stop the uncontrolled air, then add insulation and you'll be cosy.

Airtightness needs a detailed approach – don't settle for 'good enough'. This is the moment to get nit-picky, double- and triple-checking everything you do.

- Each piece of membrane has adhesive on the edge so it can be joined to the next with no gaps.
- Special, extremely sticky tape is used to link the membrane to the ceiling with no gaps.
- The same tape is used to tie the membrane to a concrete floor or link to another membrane under a suspended floor.

How to measure airtightness

The way to measure airtightness is with an air-permeability test or blow test for short.

The measure is air changes per hour (ACH), which relates to the number of times the total volume of air in the building is completely removed and replaced in one hour. If you have a draughty home, then this will be frequent – warm air sucked out through gaps in the building will be replaced with fresh cold air from outside. An airtight space will show low figures because the air is held in one place with nowhere to go. So the lower the figure, the more airtight the house.

We'll suck the air out of your house

At least that's what I thought the website said. Turns out I was being overly dramatic, imagining we'd have to stand in the garden because we wouldn't be able to breathe inside!

A blow test or airtightness test is designed to root out the places where uncontrolled air can get into the building. A massive fan is installed in a doorway to suck air into the house, making any draughts really obvious.

It was a challenge to fit the fan in our front door so it ended up on one of the double doors in our kitchen. Once it was fitted, we could start discovering the state of our build.

I picked up my trusty notebook and pen and we set off on a quest for loose air. It was all pretty alarming – even the space for the sash window ropes had a gale blowing through it. Of course it wasn't nearly that bad in reality – the fan was exaggerating every spare bit of air we had.

The list was long and the air changes per hour much higher than we'd hoped. But at least it told us what we'd missed. Like a section under the stairs that linked directly through to the sub-floor void and external air bricks. Same under the kitchen cupboards and through some of the windows.

I still have that list. Most of it is done now. Just a couple more bits of work, then we'll get another test done. Fingers crossed for a more acceptable result.

How does the blow test work?

The airtightness test or blower door test checks air leakage in a building to show you where you could improve energy efficiency. The huge fan creates a pressure difference between inside and outside. All the intentional ventilation units are blocked off – cooker hood, bathroom vent, trickle vents etc. – so they don't skew the findings. Once the fan gets going, sensors measure the airflow and pressure differences so air leaks in and out of the building can be calculated.

A new build will usually achieve between 3 and 5 ACH (air changes per hour, remember), whereas Passive House homes achieve around 0.6 ACH – basically that's no draughts at all! Remember, you want the number to be low because you don't want uncontrolled air leaking into the building.

If you're retrofitting your existing home you don't have to get a blow test done. It's not obligatory, but it's definitely to be recommended. If you're keen to make your home truly energy efficient and future-proof, then this helps you find the places that need attention.

When to do a blow test

The best time to check out your airtightness is when the envelope is completed and before you put in the significant additions to the house – i.e.:

- once the airtight membrane is fully installed, insulation is in place and you have the plasterboard or equivalent on
- before you put in kitchens, bathrooms, floors, cupboards, carpets, decorating.

In essence, do it while you still have access to the fabric of your home so you can make the changes that will improve your efficiency.

We didn't know a blow test was an option until the kitchen was in place. In fact we were just waiting to have a carpet laid in one room, so you can see we were well on our way back to normality. So it was a shock and irritation to discover we still had work to do. We've managed the bulk of it. I say 'we'. It was John lying on the floor to shove more insulation in under the kitchen units and patching up under the stairs.

Don't get stuck like us. Take stock and get your blow test as early as you can.

Quick pointers – where are the draughts?

You'd be surprised where draughts come into a building:

- **Ceiling lights:** you can get special light holders that cut out the draughts if you're aiming for a very low score.
- **Skirting board:** around the edges of a room where the skirting board meets the floor.
- **Down chimneys:** chimneys are just huge holes in your home, so the best option is to take them out if you can. If you don't want to do that, look for ways to block them off when they're not being used.
- **Doors and windows:** any place where two parts of the building come together has the possibility for air movement.
- **Pipe inlets:** plumbers will often make bigger holes than needed because it makes their job easier. Fair play, but it leaves you with a draught. Ask them to reduce the size or fill the gaps with insulation once they've finished. And check how well they've done it before they leave the job.
- **Floorboards** sitting over a suspended floor will let howling gales into the house because of the airbricks into the sub-floor void.

NB can't say it too often: never block up the airbricks.

Can a house be too airtight?

Whether a building can be too airtight all depends on your plans regarding ventilation. The two go together perfectly because:

- airtightness cuts out the uncontrolled air
- ventilation puts back in controlled air.

So if you make airtight without ventilation, then yes, the house can be too airtight and you'll have major problems with condensation. As long as you put the two elements together you'll be warm and comfortable, saving a load of money on your energy bills.

INSULATION

Then it's time to talk about insulation. There's lots more about this, including all the different options, in chapters 6 and 11.

Insulation is familiar territory for us. It gets talked about in the media, grants are given to help people add it to their homes and it gets blamed whenever condensation raises its ugly head.

It's a great partner to airtightness. You need both if you want a warm and cosy home that won't break the bank. Put in insulation without airtightness and you'll decide it doesn't work – it's all a con – because you'll still be chilly.

The aim of insulation is to increase the size of the barrier between inside and outside, reducing the loss of heat through the fabric of the house.

Typically heat loss in a home is:

- 25% through the roof
- 35% though the walls
- 15% through the floor
- 25% through windows and doors.

Insulation is one way of stopping the cold coming into the house from outside. If you get the correct materials, it is also the best way to keep a home cool in the summer.

How to insulate

You have two options: you can insulate the external walls of your home or you can add internal wall insulation.

Benefits of external wall insulation (EWI)

- Less disruption to daily life – it all happens outside.
- Makes the home much warmer.
- Less risk of cold bridges (gaps in the insulation that leave cold patches).
- Can act as a weather barrier, reducing damage to the external fabric of the building.
- No impact on living space.

Risks of EWI

- If not installed correctly, increased risks of condensation leading to damage to the wall itself.
- May need planning permission and may not be allowed if you live in a conservation area.
- Can change the appearance of a building so it no longer fits in with the area.
- High cost of scaffolding.
- Has to be done all in one go. No chance to stagger the work to fit budgets.

This is definitely a moment to find out as much as you can about an installer. Don't assume, just because they have a business providing EWI, that they understand the materials they use and the materials needed for your home. Loads of companies are doing it now without recognising the complexity involved. (See chapter 11.)

Benefits of internal wall insulation (IWI)

- You can retrofit at a rate that suits you – all at once or one room at a time.
- No planning permission needed and no concern in conservation areas.
- Reduces condensation and mould by providing a warm surface within the room.
- You can do it yourself if you enjoy DIY – and you're good at it.
- Wider choice of materials.

Risks of IWI

- You lose a little room space (although a lot less than you'd think).
- Installation will impact your daily life.
- If not properly installed it can trap moisture, leading to condensation and mould.
- You may need to replace features in your home – e.g. skirting board, coving, picture rails.

As with EWI, it's best to find someone who really understands the process, and that's not easy. However, if you're up for doing the learning, it is possible to gain an understanding in your own time.

Which materials to use

It's important to get the right materials for the outcome you want. And there are so many to choose from – far more than we are led to believe.

Builders will almost certainly default to the materials they're familiar with and the materials they can get hold of easily. They'll have an account at the local builders' merchant, where they get a good price, so they'll take what's held in stock. It does make life easier, no doubt about that – but only for the builder.

Identifying the good, the bad and the ugly

Since builders don't know about insulation options, it'll be down to you to go exploring. This is a quick summary of the measures that will help you understand what's good and what isn't.

Thermal value: this tells you how good the material is at keeping you warm in winter. Unfortunately, different providers use different measures so you may come across:

- U value
- R value
- K/Lamda value.

Decrement delay: we didn't need to bother with this until recently. But as summers get hotter we need to check – will the insulation material also keep us cooler in summer? (See chapter 6)

We're so used to the idea that insulation keeps the cold out that it's a shift to realise it can also keep the heat out. Essentially, putting in insulation creates a barrier between inside and outside so it makes sense that it can help in heat and cold. We just need that change in mindset and we'll be able to use it to best effect.

This is why you're looking for decrement delay, which tells us how long it takes for heat to travel through the material. So clearly we want the number to be a big one – the longer it takes, the better.

Thermal mass: you also want a material with high thermal mass. This is the ability of a material to absorb, store and release energy. If it takes a long time to do this, then that helps to even out the temperature in the room.

Questions to ask about insulation:

- Is it vapour permeable? Will it allow water vapour to move within the material itself, so helping to avoid condensation and possible mould?
- Will it keep me warm in winter and cool in summer? What is the decrement delay and thermal mass?
- Is the material eco-friendly – is it a natural material?
- Will it off-gas VOCs and can it be bad for the person installing it?
- How easy is it to access and how will it be delivered?

Importance of location

However you decide to insulate, you need to check on the amount of insulation your location can manage. If you live near the coast where there's a lot of moisture in the air, you need to be sure you have the right materials at the right thickness.

This is all because of the dew point. That's the temperature at which the air can no longer hold onto water vapour. Think about flowers in the very early morning. As the air gets really cold just before dawn, it can no longer hold as much vapour. The resulting moisture is what we see sitting like fairy pools in lupin leaves. The temperature at which that condensation happens is the dew point.

You need the dew point to happen outside your insulation and wall. If it happens inside the wall, there will be damage from the trapped moisture .

There are a load of calculations involved in working this all out. You can try asking your builder – you may be lucky and have found an eco specialist versed in building physics. If not, then your best bet is to ask the insulation provider to do the maths for you.* They'll be able to factor in your location, type of house and choice of insulation and tell you how much of which insulation will do the best job for you.

One small warning: if you're going to use sustainable insulation – and I hope you'll at least consider it – then you'll need to order materials in advance. There are a few local suppliers who'll order materials in for you, but you may still need to order ahead of time. That can take a few days, so think ahead to make sure you have it when you need it.

VENTILATION

Having worked so hard to warm your home up and get rid of draughts and uncontrolled air, you have to think about adding fresh air back in. I know – sounds bonkers – but we do need to breath!

While draughts are a pain, they also provide fresh air and are a real help in managing the condensation we automatically produce just from living – 20 litres a day for a family of four. Imagine living in your airtight home with all that vapour sloshing about – deeply uncomfortable.

The answer is to add fresh air back in, but this time air that you can control. Now you need air that will manage the moisture, hold onto the warmth and be filtered free of toxins, pollen and pollution. And air that goes where you need it.

Some countries take this really seriously, although remarkably few: only France, Sweden, Estonia, Germany and Finland. In terms of healthy homes (see chapter 16) this is an important part of the equation. Lack of fresh air fosters low energy, spread of infection, respiratory problems and even sick building syndrome.

Ventilation levels are measured in litres per second (l/s) – the volume of air moving through a system or opening in a given time

Building regulations set minimum ventilation rates per room (for example, 13 l/s for a kitchen and 8 l/s for a bathroom).

Some designers also use ACH (see above) as a way of comparing whole-house ventilation levels but since this is the same measure used for airtightness, it can get a bit confusing. Probably safer to stick with litres per second as the measure for ventilation and ask your designer (if you have one) to work to that.

How to change the air

There are different ways to change the air in a room or a house:

- open windows
- trickle vents
- extractor fans
- single-room heat recovery units
- mechanical ventilation for the whole house.

Opening windows

Pretty self-explanatory, but how effective is it in reality? To maintain the required standard, we'd need to open the windows fully in a room multiple times a day and how many of us are going to do this, especially in winter? A warm summer day is no problem, but in winter it's pretty unlikely – at least as much as we should.

If window opening is your only option, try this – advice given to me by a retrofit coordinator:

"Start at one end of the house and begin opening windows. Leave each window open as you move to the next room. As soon as you've finished, go back to the start and close them all." That will be a real help in keeping the house fresh and reducing condensation.

I've since discovered that this is a system used regularly in Germany. It's called Stoßlüften or "shock ventilation". It's more energy-efficient than leaving windows slightly ajar all day - it flushes out humidity and pollutants without a constant loss of heat.

Trickle vents

Builders will talk to you about trickle vents or just assume it's what you want.

Trickle vents are the open grilles at the top of your double- or triple-glazed windows. Some have covers that can be opened and closed, while some are constantly open. They allow fresh air to come into the room and are required by building regulations.

The big problem with trickle vents is that people close them up. They are essentially a draught coming into your room, so the temptation to stop cold air blowing down your neck is huge.

New windows will include trickle vents as a matter of course. If you are planning a mechanical ventilation system then you need to ask specifically for trickle vents not to be built in.

Trickle vents aside, ventilation is unlikely ever to be brought up by your builders. Don't ask me why. It should be automatic. Maybe we need an obligatory invite for builders to visit after six months to review any problems. A different, rather sweaty form of snagging.

Extractor fans

We're all very familiar with having extractor fans in bathrooms, toilets and kitchens. They serve to extract moist air and smells from busy areas. When they run constantly, they'll help with condensation by pulling out the vapour-laden air.

The problem is that many, especially in bathrooms and toilets, are connected to the light switch so only work when someone is actually in the room. If you think about having a shower or a bath, it takes a while for the moist air to disperse, so leaving the room and switching off the light is asking for mould formation. And of course, bathrooms are prime targets for mould.

If an extractor fan is your only option:

- separate the extractor from the light so it carries on working when you leave the room
- leave the light on for a time afterwards to make sure the room dries out thoroughly
- invest in an intelligent extractor where you can set humidity levels. This will ensure the extraction of moist air continues while it's needed.

All these options will make a difference to moisture levels and help reduce the occurrence of mould.

Single-room heat recovery units

We are far out of builder territory here. Even architects and other professionals are not always up to speed with single-room units. My mentor for *Beginner's Guide to Eco Renovation* said this is one of the things I taught her!

All you need to install a single-room heat recovery unit is an outside wall, which makes it a viable option, especially for those who want to retrofit one room at a time.

The units work by transferring heat from the outgoing air to the incoming air. This happens in the part of the unit that sits in the wall and it holds onto the heat that costs you so much to create. The net effect is that you don't notice them working. Unlike standard air vents, where you can feel the draught as the air moves in and out of the house, the air coming in is at pretty much the same temperature as the room.

At the steady state level, they are also very quiet – unlike the usual bathroom extractor fan. They are noticeable when humidity levels rise and the fan switches up to a strong level to deal with it. But that is short-term. Overall, you're unlikely to notice them*.

Did you know we need ventilation?

Imagine the situation: builders already hard at work, cold house with few windows and doors in place, just a bedroom and one living space to house me, John and regular visits from our three-year-old grandson (sometimes accompanied by the two dogs). The dust was lying thick, we both looked like the Michelin man wrapped up in most of the clothes we owned. Daily, I wondered why on earth we ever thought this was a good idea.

My laptop was my constant companion – my bit of sanity and my teacher. As ever, I'd been researching what needed to happen when retrofitting and I'd stumbled over a piece about ventilation. No one had mentioned this – I guess the architect didn't see the need. Our new extension, designed by him, would be airtight but the rest of the house was mostly ventilation.

But, thank heaven, I persisted. Part of my stumbling around the web brought me to the single-room heat recovery unit. Designed to manage one room, it had three speeds which adjusted according to the humidity level we set on an app. And the best bit? It had a 'party mode'. I liked their thinking!

The heat recovery part was a huge draw. Remember we were retrofitting to create a legacy for the future, so anything that held onto the heat we'd create had to be a good option. The blurb claims 85% of the heat is retained, passing from the warm outgoing air to the cold incoming air.

All we needed was an outside wall – and we could do that. At that point we thought one in the bathroom and one in the kitchen would be fine. So after much discussion and many calls to the tech people at the company, I set about ordering a midi for the bathroom and a maxi for the kitchen/dining room area.

Four years later we have one in each room. We soon understood that our need was greater than just two units. Especially given that we made the home as airtight as we could. Having got rid of the natural Victorian ventilation system, we needed some controlled air to make up the shortfall. And the beauty of the single units is that we could add them in as we realised the need.

Single-room heat recovery units can be installed without all the disruption of putting in a whole-house system. The only possible hitch comes if you live in a conservation area or a complex where they don't allow anything to go on the outside front wall. We had the conservation challenge but managed to install a unit in our front room by venting it into the open front porch. Otherwise there is nothing to stop you installing what you need.

How does heat recovery work on hot days?

It all makes total sense when we think about keeping the house warm in winter. But what about super-hot days in summer – then we want inside to stay cool. There are still ways that the units can help keep the home cool:

At night: switch the units to 'supply only' so they bring in fresher night air and cool the whole house down. It'll be easier to sleep and will set you up for the next day.

By day: close the house down and switch the units back to normal 'heat recovery' mode. The ceramic core of the unit responds to the temperature inside the room, so as long as you have cooled the house effectively overnight, it will cool the incoming air to the same temperature.

Managing summer humidity: when the air is very humid outside, the unit will take a little longer to dry the air in your kitchen after cooking or your bathroom after showering. This is because the moisture condenses on the ceramic core, slowing up the functioning of the unit. So don't panic – like we did – it's not broken, just responding to the humidity. It will get there in the end.

If you have a major problem with condensation, you have the choice to leave it on supply only for a time until the air begins to dry enough to revert to heat recovery.

Some questions to ask about single-room heat recovery units:

- How often do filters need changing? Can I do it myself and how much are they?

- Which unit will be best for the size and function of my room?
- Is there a remote control for the unit? (This is helpful for a quick change of speed.)
- How do I adjust humidity levels? Is there an app for my phone?
- How much does each unit cost and who does the fitting?

Mechanical ventilation heat recovery (MVHR)

This is the crème de la crème of ventilation systems so install this if you can. The main unit sits in your loft with ducting that feeds vents in each room. If you live in a flat you may find there is already an MVHR that serves the whole block. If not, you could talk with your neighbours about fitting one.

Not only does the system bring you fresh filtered air, it switches heat between outgoing and incoming air within the unit. In the process the filters remove pollutants like pollen, dust and mould spores. The filters are changed on a regular basis to ensure the air quality in your home remains high.

MVHR gets more sophisticated every day, so do some research to make sure you're looking at the most up-to-date models*. There are also some now that can sit outside the house, removing the need to have space in your loft.

Check out filters and what they remove from the air. This too is advancing fast so if you have allergies or respiratory problems this could be a real game changer for you.

Questions to ask the experts when exploring MVHR:

- How often do filters need changing and how much do they cost?

- Can I change/clean the filters myself or do I have to pay for it to be done?
- Is there a summer bypass mode or a way to cool the house when needed?
- Does the system we're looking at have the option for a small cooling unit? (This brings cooled water to a coil in the fresh air duct that reduces the temperature of incoming air.)
- Can it be connected to my heat pump? Some systems can be linked to reversible heat pumps that provide active cooling – more efficient than summer bypass on its own.

"I'm renovating my 1880 lime-mortar house"

One of the joys of Instagram is the people who get in touch to tell me about the work they're doing to their homes. So when Ben dropped me a note about his 1880 house, I was thrilled. Even better, he wasn't that far away from me so I could go and visit.

A fantastic building in lovely grounds, it had been the home of a small group of nuns – and that's a story you don't hear often! It's a breathable solid brick structure, so Ben had loads of work to do putting right the damage caused by uninformed builders. Bricks were crumbling from the onslaught of gypsum plaster, damp-proof course and plasterboard.

When I first visited, the walls were back to the brick and he was coming towards the end of installing an MVHR system. Doing it himself, too – which I found very impressive*.

It's quite unusual to see MVHR going into an old house because the ducting that needs to go between floors is bigger than the space will normally allow.

However, Ben had found a system with rectangular ducting that was quite flat, so took up much less room.

I was really excited to see MVHR becoming an option in a solid brick build – it provides improved air filtration but also the option of air cooling in our increasingly hot summers.

The most important message I want you to take away from this section is that you **must** take positive action to add controlled air into your retrofitted home. If this isn't done, condensation will become an issue. The sad thing is that it's often insulation that takes the blame when, in fact, it's just the lack of controlled air.

BREATHABILITY

Let's talk breathability, because it's basic to the issue of condensation – the scourge of 21st century living.

It's a really daft descriptor, because it's nothing to do with breathing and everything to do with water vapour. Much better to talk about vapour permeable or vapour open. However, I'm going to use both in this book so you're familiar with the term if it arises. I don't want you feeling befuddled just when you need to make sense.

There is one positive. Because it's such a silly label, it's a great way to sort out those who understand retrofit from those who just claim to. I slip the word into conversation to see how people respond. The uninformed will immediately wax lyrical about air flowing all over the place. That's the moment to move on!

Vapour permeable means the material has a moisture pathway that allows water vapour molecules to pass through. We're not talking huge amounts of water, but enough to make a difference.

Get to grips with this before the builders come in or even before asking for estimates. It's something to discuss with your architect, if you decide to work with one, because its fundamental.

If a builder arrives at an old breathable house armed with PIR insulation or gypsum plaster (see chapter 6) you have a big problem. Use a plastic-based, synthetic material on a breathable wall and the chances of managing vapour are minimal to none:

- if it's installed inside your house, vapour hits the impermeable barrier, can go no further and condenses on the wall
- if it's installed on an external wall, vapour passes through the wall, meets the barrier and condenses within the wall.

Internally, you can see what's happening and take action; externally, you don't know anything about it until you have major problems with bricks crumbling and breaking down.

New builds: a newer house is automatically non-permeable, but breathability is still an asset. While vapour can't have free flow, it will still be absorbed by the vapour-open material and held until the air is dry, at which point it will be released back into the room.

Even the newest of builds can have a problem with vapour. If you are in that situation, read about ventilation again and consider installing breathable materials: insulation, plaster and paint. Take a look at chapter 11 and read about clay plaster. If you have one room that is particularly prone to vapour or condensation, then taking the plaster off and redoing it with clay could be a really good option.

Summary

- The big five of retrofit form the backbone of your energy-efficient home: airtightness, insulation, ventilation, sustainable heating, renewable energy.
- Breathability is an important element for those in older homes. Miss this one out and you increase your risk of condensation and mould.
- Ventilation is the least talked about and most important element of energy efficiency. When you cut out draughts you must add back in controlled air.
- You can insulate externally or internally. Both work well as long as you have the right materials for your home.
- All insulation will make your home warmer in winter. Not all insulation will make your home cool in summer. Look for materials that have a long decrement delay and high thermal mass. **NB:** the builder won't tell you about this.

To find the information indicated by the * follow the QR code.

NOTES

CHAPTER 11
ALL YOU NEED TO KNOW ABOUT NATURAL MATERIALS

I feel very rude suggesting that builders don't know about alternative materials, but my experience to date is that it's true. Of course there are some in the know, who use different materials on a daily basis, but you have to look hard to find them. Just now, anyway.

This is an exciting time to be building because we're at the beginning of a major shift. For the sake of the planet and future generations it's time to move away from synthetic materials that aren't good for our health (VOCs and lack of permeability leading to increased condensation) or for the environment (petrochemicals/gas releasing significant carbon into the atmosphere and changing the climate).

On the whole, human beings are creatures of habit. Some people absolutely love change, but others prefer to stick with what they know – they just distrust the unfamiliar. The most common response to change is resistance. Just post something on social media about reducing meat consumption and you'll see what I mean. It's as if you've threatened life itself!

When it comes to building and renovating, there are a whole range of materials that can make a major difference to the comfort and effectiveness of your home, as well as saving you a shedload of money on bills. We just need to be open to thinking about them.

Builders might seem the obvious place to start, but that means asking them to change what – to their mind – already works fine. Why would they bother? It's all going great, thank you very much.

There are a few reasons for this:

- builders don't know there's a shift to be made and most have never heard about the different options outlined here
- life has become so high-tech and modern that there's resistance to hearing about natural materials and an assumption that they just won't work
- there's also a belief that building regulations may make it harder to get work approved if it's done in a 'strange' way. (Hemp is now accepted by Ofgem - the energy regulator)*.

It's interesting that in our sophisticated, high-tech world the materials that offer most help in the climate crisis are those that have been used for centuries. The Egyptians built with clay and lime, as did the Romans. In fact, the Romans even had concrete. It was made from volcanic ash, lime and water and had the ability to gain strength over time. We can see just how well it did in Rome – so much is left standing from over 2,000 years ago, which is a lot better than we do right now.

Houses are built with straw bales and definitely don't get blown down by the big bad wolf! Cob in all its many forms – essentially mud, clay, straw and lime – is used to build homesRem in Ireland, Bulgaria, the UK, France, the Middle East, Africa, Southwest United States. And wood-frame houses are coming back into fashion, based on amazing builds in the US and early England, where they were paired with wattle and daub (clay, plaster and animal dung!). I think I'd be with the builders about the animal dung, but otherwise it sounds good to me.

How do we make this happen?

Never underestimate the challenge of trying something new, especially when it affects your work. Builders want to do a good job for you; they want you to have what you want and using something unfamiliar can just feel too much of a risk. If you want something different then it's for you to share your ideas with knowledge and confidence.

The key is finding builders who are either expert in the use of natural materials or interested and willing to learn.

Working with clay or lime, for example, will need an eco builder, maybe even heritage builders if they are willing to work in more modern buildings. You need to discover ahead of time whether they're interested and knowledgeable in the area you need. Possible questions might be:

- How do they bring sustainability into their work?
- What materials are they used to using?
- Are they interested and willing to learn about other options?
- What eco projects have they worked on and can you speak with the client?
- Why is eco building important to them?

My own way of telling is to ask about breathability in relation to a specific material for my Victorian home. This is the killer question and relies on your knowing something about it yourself, otherwise there is no benefit.

It's useful because 'breathability' is such a strange choice of word in the building dictionary. Remember, it actually refers to vapour permeability, so is about moisture (see chapter 10). Those familiar with sustainable materials and building will probably point out that you're speaking about movement of moisture. Those who don't know are more likely to assure you that air can move easily in the material.

Is it breathable?

I gave a talk a while ago about retrofit to a very keen and active group of homeowners. Present in the audience and clearly wanting to make connections with possible customers was a provider who installed external wall insulation (EWI).

I had made the point in my talk – I always do – that older homes (pre-1930) need vapour-permeable materials to avoid condensation and mould. As I was leaving, I was pleased to hear someone talking about the possibility of EWI and asking about breathability – I'd clearly got my point across – but they were being brushed off.

It had been a busy session and I was ready for a cuppa at home, so I started back to the car. But I couldn't do it. I had to go back to find out more.

I approached the provider and asked about the materials they'd use on a Victorian home – there are loads in my home town, so it was a relevant question. He assured me all the materials used were breathable. Sounded great, but then he broke the spell by explaining to me that they were polystyrene-based so air could move easily. And it had been going so well!

Moisture is a constant problem in modern building, a problem that is rarely an issue in an older build – unless modern products have been introduced inappropriately.

Natural materials hold moisture and even balance humidity in the room, so they work well with human beings. We're not built to be ultra-dry with no fresh air, and we produce moisture all on our own that has to be dealt with.

Synthetic products are closed-cell structures, which means they can't wick moisture away from the living space. End result? That 20 litres produced by a family is sloshing around with nowhere to go other than dripping down the walls.

In contrast, natural products – hemp, wood fibre, straw, sheep's wool, clay, lime – are all open-cell, so manage the water vapour before it builds up in the house. Take out the problem of moisture and we more than halve the problems we experience in keeping our homes warm and in good condition.

BUILDING WITH NATURAL MATERIALS

Options for natural building:

- straw bale houses
- screw piles
- glass aggregate/glass beads
- hempcrete
- tyrecrete
- clay bricks.

Straw bale houses

Imagine being able to say to your friends that you live in a straw house. First, they won't believe you, then they'll question your sanity. All because it's a way of building that we're unfamiliar with.

There are distinct advantages to building with straw:

- you will stay warm in winter and cool in summer
- the house will last for at least 100 years
- straw can be local so you reduce carbon emissions.

Because it is unusual, it can take a long time to get through the design and planning phase, which is daunting, but once that's done it won't take any longer than a normal build. In fact the structure of the house itself, once foundations are in place, will be done very quickly.*

It's perfect from a sustainability point of view – we produce enough straw in the UK to build 600,000 houses, and build just 100,000. Think of all the concrete, cement and bricks we'd save by building with straw.

Everything used with a straw bale needs to be vapour permeable, so the ideal plaster is lime or even clay. Both are breathable, both add additional insulation and both add to the healthy home by purifying the air.

Two common concerns:

- **Mice and rats:** since straw is the empty stem of the grain crop it doesn't contain any food, so there's nothing to attract vermin. Leaving crumbs out in your kitchen will cause a lot more problems, than the build of your house. And once your straw bale building is plastered, the walls are the same as any other wall to a rat or mouse.

- **Fire:** when you stack bales and plaster them, the density is such that there isn't enough air in the bales to burn. Straw bale walls have passed all the fire tests they have been subjected to in Europe, the USA and Canada. And, regardless of whether the bales themselves are a risk, if you plaster any wall with a half inch of plaster, it gives sufficient fire protection to satisfy building regulations. In fact, plastered straw bale walls are so much <u>not</u> a risk that they are being used as fire walls between semi-detached houses.

Screw piles

Every home or extension needs good foundations. This generally involves a great deal of concrete being poured – 450mm to 1m or more, depending on the earth you're digging into. That's a lot of concrete and a lot of emissions.

But there is an option – screw piles* are much lower in emissions but also lower in price. They are installed about 1.5m apart and a base is constructed on top of them. It's quicker too, since there is no drying time. As soon as the screws are in place the next piece of work can start.

I've never seen anything like it

Building our garden barn, we were determined to be as eco-friendly as possible. If something wasn't good for the planet, it wasn't included.

So the day they came to install the screw piles was very exciting. The screws themselves came up to my shoulder, and I'm 5'8" – they are huge! Two men came to install them. Once they'd sorted out the plot, they began.

The first stage was done by hand as they got the screw to bite into the earth. Then they had a machine that continued the process. But – and this was the best bit – to add extra heft one of the guys got on top

of the screw and sat on it as it pushed down into the ground. Follow the QR to see a reel on my Instagram account.*

The end result is good quality underpinning for the building, considerably less disturbance to the earth, next to no emissions – and a lot of fun!

Compacted hardcore / 6F2

Some projects will need foundations that are more robust, to take the weight of a bigger building. Usually this would be a trench or hole full of solid concrete – gives me the shivers just to think of it – but there are alternatives. 6F2 is made from recycled construction and demolition waste, so at least it's not using brand new materials. It's made up of crushed concrete, bricks and mortar. It's commonly used as the sub-base for roads, driveways, patios and building foundations.

So even for the most demanding elements of a build, we can find a way to reuse and repurpose materials to reduce cost in money and carbon.

Timber frame: this is an eco-friendly way to build a house and, with a quarter of all new builds being timber frame, it's growing in popularity. Wood is sustainable with low embodied carbon, not to mention that it holds onto carbon rather than releasing it into the atmosphere.

A new advance with timber frame is to begin construction off site. The frame can be started, then put up like wooden Meccano and the infill part of the construction can be added in the form of structural insulated panels (SIPS)*. These are panels that are built to size to fit directly into the gaps in the frame and they come already airtight and insulated.

Because SIPS are made to fit, you can get them filled with a number of different forms of insulation. The most commonly used are, of course, PIR (polyurethane), expanded polystyrene (EPS) and graphite polystyrene (GPS). These materials will be sandwiched between two layers of OSD (oriented strand boards) to create a strong, well-insulated panel. (See chapter 6.)

You can, however, have SIPS made that use very different insulation materials. The concept is the same – panels made to suit, fully insulated and ready to drop into place. The only difference is the insulation materials used in the panels. Warmcell uses cellulose which is made from recycled newspaper and Bio-SIP makes the most of recycled plastic bottles.*

Foam Glass Aggregate

In our modern world, when it comes to creating the floor of an extension or house, the standard option is usually block and beam. This is a suspended floor but all made with concrete beams and blocks with an air gap underneath. But again there are options.

Foam glass aggregate (FGA) can provide the structural support you need with some insulation thrown in. It's made from recycled glass, although it looks nothing like glass, more like coal clinker. It's distributed over the earth, then compacted down with a whacker (the actual technical term!), then the top layer is added.

Usually that top layer would be concrete, but again there are options. Hempcrete and glass beads*, for example, will give you the floor you want with added insulation.

Hempcrete

Anything to do with hemp is amazing. It's a remarkable plant that grows so fast it can be harvested twice a year, which means it can sequester loads of carbon, tying it up for the foreseeable. Not to mention that it's a great insulator with a good thermal resistance, which means you'll stay cool in summer as well as be warm in winter.*

Hempcrete uses the shiv of the hemp plant – the chopped woody core or stalk of the industrial hemp plant, left after the seeds are removed. It's mixed with lime and water to make solid bricks. The beauty is that it is breathable (vapour permeable) and provides really good insulation, as well as exceptional carbon capture.

It was not so long ago that I saw my first hempcrete block. James King from Unity Lime* sent me one for my box of insulation samples. I was so excited – embarrassingly so, if I'm honest. But it's fantastic to find a material that is good for the planet and good for us. Now I can show it to all the folk who visit my Open House sessions. I lay out the contents of my box – wood fibre, denim, diathonite and now hempcrete – so they can see and feel all the materials we're talking about. It's much easier to talk with a builder when you've seen it, held it, understand what it's used for.

Tyrecrete

This is an entirely new one to me and only just on the market. I've seen a sample and spoken with the inventor, but I'm not sure how available it is yet. Since you may be reading this well after publication, I'll include it here so you can go searching.

It's not an eco material in itself, but it is a good use of old rubber car tyres, which are usually dumped or sent overseas to be burned once they've reach the end of their useful life.

The idea is that the rubber from waste tyres is shredded and used to replace 5% of the aggregate in manufacture of concrete. The benefits for the material are that it is lighter, so the cost of transport is lower, and it also makes the concrete more flexible. Flexibility is useful for bridges, earthquake areas, water infrastructure. Not quite sure why it would be helpful in a home, unless you live near a railway line that produces vibrations maybe?

At least you know it's an option!

Clay bricks

The final building option is to use clay bricks. Clay, like hemp, is an amazing material that serves us in many different ways. People have built with clay for thousands of years, the earliest known use being in 7,500 BC. As a testament to its durability, we can still look at the Great Wall of China, the Red Fort in Delhi and Hampton Court in the UK.

Most bricks used today are made of clay. The big difference between a brick bought at the builders' yard and traditional clay bricks is the use of heat. Standard bricks are heated to 1,000°C, whereas clay bricks are heated to 50°C. The benefit of low heat is that clay bricks can be put back into water, returning them to raw clay so they can be used again, whereas fired bricks will only ever be bricks or broken up into rubble.

Clay bricks have a number of advantages in terms of a healthy home. Clay manages your moisture for you by absorbing vapour when humidity is high, then releasing it back into the room as the air dries out. The clay also has no VOCs in it, so you get a cleaner atmosphere.

If you're interested in earth building you might like to explore Strocks*. These are structural blocks of clay-rich earth and chopped straw, which can be used as the inner skin of an external wall and for internal load-bearing walls, typically up to three storeys. The beauty is they come in the form of a block, so they're easier to build with than having to get in among a bucket of clay.

Unfired clay has extremely low embodied carbon and it's 100% natural, fully recyclable and quick and easy to use.

We built a south-facing roof

Once our retrofit had settled and we were ready for building again, we took down our existing garden office and rebuilt so it had a south-facing roof for solar panels.

By this time we had learned a great deal about the different building options, having worked our way through the retrofit (with minimal support) and I'd written a book (with loads of support) so we were really excited about the chance to put it all into practice.

We began with screw piles for the foundation. Driven by a determination not to use concrete, which is very water greedy and energy greedy in production, this also suited the needs of a sloping garden.

The end of the new build was going to sit on the foundation left from the old office – no point in lifting concrete just to throw it on the dump – but it wasn't quite high enough. This was our chance to try blown glass beads*. That was great fun!

The barn is timber frame, lined with an airtightness membrane, 100mm of flexible wood fibre, 80mm of rigid wood fibre and finally wood wool board as the surface to take the breathable paint. We decided against plastering because we liked the texture of the wood wool board.

The roof and back of the building are sheet metal, needed for fire resistance. The front and side that's on view are clad with a beautiful European larch.

Never ones to do anything the easy way, we found a master carpenter to work with us to put up the frame.

He'd been at it for many years and now likes to work alone, focusing on the structure and creating the most incredible joints. It took a long time, but we were in no hurry. As soon as the roof was up the panels could be installed and that was the main purpose.

We learned a lot. Not least the delight of using recycled materials. A second-hand triple-glazed window from a local recycled window company; many of the timbers for the framework – include my clothesline prop (twice) – and a 70-year-old gym floor made of solid beech. Using preloved brings its own challenges but with a determination to be as eco-friendly as we could, the satisfaction now makes it all worthwhile.

INSULATION

This is so exciting! I know – building bore alert!

There are a number of fascinating options for insulation, but synthetic petrochemical-based materials are the default because we don't know we have a choice. And it's an unusual builder who will suggest something different.

If I'd been better informed when we were planning our extension, it would be a very different beast. I'd want clay bricks, hemp insulation, and foam glass aggregate for the floor topped with limecrete or glass beads. And of course, I'd have had a combination of clay and lime plaster.

We can only work with what we know and all we knew about at the time was wood fibre. It is a great insulation and the house is working extremely well – saving 75% of energy in a 1901 Victorian house is pretty spectacular. It's just fun to think about all the options we could have gone for.

When it comes to what insulation to use, there are three things to take into account:

- matching the material to your house
- ease of access
- what the material will deliver.

Matching the material to your home

Finding the right match for your home will depend on two things:

- how old your home is
- how environmentally friendly you want to be.

Any home built before 1930 and some built between 1930 and 1940 will be breathable (vapour permeable).

Other signs that you have a breathable home are that it will have a suspended floor over a sub-floor void and it will be a solid brick build. Another clue will be the way the bricks are arranged in the wall. Old houses tend to use Flemish bond, which means bricks are placed at right angles to each other*.

If you're working your way through the book from the beginning you'll know this already, but just in case you're dipping in: older houses as described above were built with materials that are vapour permeable. In short, this means that any vapour produced in your home can move easily through the walls. As long as vapour can move, your risk of condensation is minimal to none. Block that movement and the risks increase hugely, with a good chance that mould is on your horizon. (See chapter 10).

If you live in a new build (after 1940) breathability is unlikely to be an issue. This was when modern materials were entering the marketplace. The cavity wall was coming into its own, with the air gap providing the only insulation included in the building.

In this set-up, synthetic materials are OK. However, air flow and condensation must still be addressed. Modern insulation will leave you living in a plastic box that holds onto moisture and, if it has nowhere to go, you know what will happen. The answer then is to install a quality ventilation system that manages humidity for you (see chapter 10). There are loads of good reasons to do that anyway, so it won't be a wasted investment.

If you want to be as environmentally friendly, then natural materials are best in your home regardless of age.

Natural insulation materials

Numerous options are available. It's surprising that we haven't heard of them and great that we can discover them now:

- hemp fibre
- wood fibre
- cellulose
- sisal
- denim
- sheep's wool
- diathonite.

Hemp fibre

This is the one that got away for me. I love the idea of using hemp. It's grown in the UK and works exceptionally well as an insulation material, keeping you warm in winter and cool in summer. This is because it has a high thermal conductivity and long decrement delay (see chapter 6), which means it takes a long time for heat to pass through the material. Winter heat stays inside the house for longer and summer heat stays outside for longer.

Not only will it keep your house at a steady state heat-wise, it's naturally fire-resistant – it will char without burning – and the material naturally resists mould, bacteria and pests like rodents. Not to mention that its acoustic properties make it ideal for drum-playing teenagers!

One of the problems for builders is that hemp is difficult to cut – unless you have a special hemp saw. So if they complain you can show off your knowledge by suggesting they contact Indi-Nature* (the prime supplier of hemp materials) about a hemp saw.

Carbon sink: I really like this bit. Hemp holds a lot of carbon in its structure. Use hemp insulation and you're giving a home to carbon sequestered from the air, meaning you keep carbon out of the atmosphere while keeping yourself warm. Brilliant!

Wood fibre

We used wood fibre in the bulk of our 1901 Victorian house retrofit. It comes in different forms – flexible insulation batts and plasterboard equivalent. The batts can be cut easily to fit the spaces required and it's a simple matter to pull bits off to fill any gaps that have been left*.

Initially I was concerned about cutting down trees to make the materials. Then I discovered it's all waste softwood fibres processed into a pulp or dry fibres that are formed into the boards or batts. As soon as I hear that waste is being used, I'm on board!

Have they done it right?

Our new extension – a room built on the footprint of the old conservatory – was designed by an eco-archi-

tect – a compromise between me (likes doors that close to cut down energy use) and my husband John (likes snazzy-looking open-plan spaces). The extension structure was: external brick wall, air gap, wood framework, airtight membrane, flexible wood fibre batts covered with rigid wood fibre 'plasterboard', lime plaster and breathable paint. Phew – I feel exhausted just remembering!

We lived in during the whole process. It was a mess, but I don't regret it. None of us really knew what we were doing – us or the builders – so living in meant we could keep a good handle on what was happening.

This is how I know that wood fibre is so flexible – it's easy to pull a bit off to fill a gap. We spent hours filling in gaps around the central heating pipes that took the hot water upstairs. It was addictive at worst, meditative at best. But all good for the airtightness.

Wood fibre is also made into rigid wood fibre insulation boards that go over the flexible insulation and provide the ideal base for plaster. This is the natural equivalent of plasterboard or drywall.

Cellulose

Cellulose is made from recycled paper products like newspapers and cardboard, so is environmentally friendly. It then has to be treated with boric acid or similar chemicals for fire, mould and insect resistance. Because of this the builder has to wear protective clothing during installation, to protect from the dust, but otherwise it's safe for air quality and biodegradable. And all made from recycled material.

Sisal

This is a very interesting product. Sisal is a fibrous plant grown in warm, moist tropical climates. Once mixed with other fibres like wool and sacking it becomes an excellent thermal and acoustic insulation.

There is one company in Scotland* that mixes virgin sisal with recycled sisal from old coffee sacks and waste wool from the carpet and Harris Tweed industry, which is rather pleasing. I love to hear about ingenious schemes like this that use every bit of waste they can get their hands on. They also make a point of saying that the sisal is grown by small-scale organic farmers in Tanzania.

Sisal is the new kid on the block in terms of its use as insulation and may well become more significant since it thrives in dry climates, so cuts out the need for additional water. You can get it in a roll or in batts and it can be used in the loft, walls and under floors.

One benefit is that it still retains its insulation properties if it gets wet, so you don't have to pull it all out if you have a leak. And no need for protective clothing when it's being installed, which is a win for you and the builders, non-toxic, no VOCs and totally breathable so suitable for any age of building. Well worth a look.

Denim

Where did that come from? Denim? For insulation?

I love this one because it uses our old denim jeans, coupled with old cotton T-shirts and velvet.* Made entirely with recycled fabric, this insulation material is lovely to work with. It's soft and non-toxic. It's important it's not compressed, because that will push the air out of it and it's the air that makes it so effective. It will lose its value if it gets wet, so it's important to include a moisture barrier with installation.

I've only ever seen one brand for denim insulation: Pavatextil, but go exploring and see if you can find others.

Sheep's wool

Many people will immediately back away from this as an option, because of the *Grand Designs* house that used sheep's wool insulation and ended up with a massive infestation of moths. What a horrendous experience!

However, don't let this put you off. You just need to know some basic information before choosing the actual material.* Many brands of sheep's wool insulation are treated with chemicals like borax. This is a fire retardant which is also toxic to the moth larvae. Another option is Permethrin, which is a chemical compound that kills the larvae but is not eco-friendly and is even restricted in some areas.

Borax can damage the fibres because it's highly alkaline and I understand it comes off as the material is moved around and during installation, all of which means it ends up less protected once in situ. Also not great having chemicals like this in your house.

Fortunately now there is a different option that is far more effective and environmentally friendly. The molecular protein structure of the wool fibre is altered by a plasma-ion treatment which stops the wool having any nutritional value. Put simply, there's nothing for the moth larvae to eat so your insulation is safe. As far as I know there is only one brand that does this at time of writing, but go search on plasma-ion treatment to see if you can find alternatives.

Sheep's wool carries all the same wonderful properties of natural insulation: it keeps you warm, purifies the air and regulates moisture. I know that fire is always a concern with a product like this, but in reality the wool chars at the first point of contact and that charring protects the rest of the wool from the heat. It's like having a scab over the surface.*

(For an interesting film of fire resistance in natural materials go to Unity Lime on YouTube - follow the QR code).

Diathonite

We used diathonite in our front room to accommodate the bay window more easily. We didn't want to risk losing the shape – it's a fundamental part of a Victorian home – so I went searching for a way to insulate and keep the window as is.

Finally I was pointed towards diathonite by a kind architect who specialises in retrofitting old houses. It's a thermal plaster, made from lime, clay, cork and diatomaceous earth (don't ask me – no idea, go search!) and it can be put on to any thickness. It ranges from 40mm to 100mm on our wall according to where it is.

Given that it's a plaster, it can be used in different ways. We also have it as a skim on the bedroom wall, which needed levelling before adding 80mm of rigid wood fibre. Why miss out on a bit of thermal value? By making sure it came right to the edge and then sealing it in, we also gained an airtightness layer. (See chapter 10.)

One final use for diathonite is as external wall insulation – same product in a slightly different form. It's perfect as an external render for older, vapour-permeable buildings. In contrast, those who use a synthetic external wall insulation find it very hard to stay cool in summer.

Because diathonite is a plaster, no definitive stats can be given for its thermal capacity because it depends on the thickness applied. But done in the right way it provides good thermal resistance so the home will be warm in winter and cool in summer, which is what we all need as the climate changes.

How do I know what I'm buying?

If you're keen to buy only materials that are environmentally friendly – and I'm sure you are since you're reading this book – take a look at the natureplus Eco label.* It's the only European environmental label for building products that's based on strict scientific criteria:

- **Sustainability of resources:** Only building products made from renewable resources, mineral raw materials that are available in abundance or secondary raw materials, are permissible for a natureplus certification. The use of fossil raw materials must be avoided whenever possible. The raw materials must stem from sustainable sources.
- **Clean and efficient production:** On-site inspections of the manufacturing facilities are conducted to verify that the manufacture of the building products is energy efficient, that the least possible burden is placed on the global climate and the environment, and that social responsibility standards are met. The products must be functional and recyclable.
- **Protection of the environment and people's health:** Building products with the natureplus label don't adversely affect the environment or human health through harmful substances and ensure healthy indoor living.

FINALLY THERE IS PLASTER

There is one final piece left in the insulation puzzle and that's plaster.

I've been finding out more about plaster recently and it's remarkably interesting. I always thought it was just the final coat to give you something to paint on. It is also an exciting moment where you finally see the room you've been working so hard to create. Suddenly it no longer looks like a building site.

Being a plasterer must be a great job. Like the hairdresser who manages to transform your look in an hour, so the plasterer transforms your space into something liveable. They must get loads of appreciation!

Plaster options

Your builder will assume you'll have standard gypsum plaster on your walls. But stop and think – don't be rushed. When you go for natural materials the plaster will do so much more than just give you nice smooth walls to paint on.

Gypsum plaster is quick setting and generally crack resistant so you can see why it's become popular. It can be done in one day and can be got out of the way so the next stage of the project can begin.

It's fire-resistant because it contains a lot of water, plus it gives that nice smooth finish we've all got so used to. It's the most commonly used plaster by standard builders and it will just be assumed that's what you'll get.

You do have options

Don't expect your builder to be much help with this. They'll almost certainly have heard of lime plaster but may not know about its value in your home. It's less likely they'll know about clay plaster, but both are really good options if you want a healthy home.

If you're plastering in an older home and/or on top of natural insulation materials, you need lime or clay. Remember, one of the big attractions of natural materials is that they're vapour permeable. Putting a non-permeable plaster on top of them means you lose all the benefit and you're back to worrying about condensation. Once vapour permeable, always vapour permeable.

Lime plaster

No one can say that lime is new-fangled. Vitruvius wrote about it in 27 BC and it's been used ever since. It's not only breathable – it actively attracts moisture then releases it back into the room as the air dries. And it's antibacterial so it purifies the air.

It's made up in different ways, depending on the needs of each job:

- Horsehair or hemp fibre are added to give strength and flexibility. This also makes it more durable and less likely to crumble or deteriorate over time.
- Its naturally high pH means mould and mildew hate it so just won't grow, which is a huge advantage in a damp climate.
- It's water resistant, so it's great for external surfaces, in basements and in the parts of a bathroom where direct splashes are likely.

And here's a nice one – it absorbs CO_2 as it cures, so it manages to absorb back all the carbon released in manufacture.

We have lime in our house and it's fine. If I didn't know what it was I'm not sure I'd have noticed any difference. It is very slightly rougher but I think that's to do with the application rather than the material. The only downside of lime is that it's caustic, so great care needs to be taken not to get it on the skin during application.

The biggest challenge we experienced was in finding a lime plasterer. They are like hen's teeth. When we finally got round to retrofitting the last room in our house – the bedroom – the first thing I did was look for the plasterer. So my advice – don't hang around, start looking as soon as you can. This isn't someone you can search for and get in the next day.

Clay plaster

I'd love to have clay somewhere in the house. Don't worry – I'm working on it*!

It has so many qualities that will make a real difference to the wellbeing of your home. For example, moisture management. Put 12mm of clay plaster in your bathroom and you'll have no fogging up of mirrors. The clay will absorb the steam from your shower, then when the room is dry enough, it will desorb the moisture back into the air. It holds the humidity of a room between 50% and 60%, which is the sweet spot for humans.

This is a real game changer for people with respiratory problems that are triggered by high humidity.

High humidity – solved

I heard recently about someone who moved into a flat with high humidity. They got hold of humidity monitors and found it was sitting at +80%, which is extremely uncomfortable.

Having tried every other option – dehumidifiers, opening windows – they set out to put clay plaster on all the walls. Sticking to the required 12mm depth, they worked through each room of the flat.

The end result? The humidity monitors were soon giving a very different reading. Once it had settled, the moisture in the flat was sitting between 50 and 60% consistently. And the only difference? The clay plaster.

You'll get no VOCs off clay, so air quality will be good and it has low embodied carbon because it's naturally occurring without the need for deep drilling to get to it. A very neat point is that it can be used again. Take the clay and put it in a bucket of water and eventually it will go back to its natural state. A truly circular product.

Another lovely feature of clay plaster is that it can be repaired. If a chunk is taken out of your wall by a sudden attack of a toddler scooter or cag-handed furniture moving, it can be managed. Add some moisture and pop on a small amount to even the area out. Perfect!

Once final point to share with your builder. No toxins or nasty products needed, so these options are much safer. All they'll need to do is run their hands under the tap at the end of the day. No protection needed, no solvents to get rid of it, just good clean washable clay.

Don't forget the paint

This is the moment you've been waiting for - finally you're on to beautifying. Any colour you want is out there somewhere and you can go ahead with creating the room of your dreams.

But stop - take a beat. Think it through for a moment.

Natural materials will give you a toxin free, vapour permeable wall build up designed to keep you dry and warm. The last thing you want is to stick a non breathable, plastic based paint of top, undoing all your good work. But don't worry, there are a growing number of paint brands that are very breathable, with low VOCs and no toxins*.

All paints are breathable to a degree, you just need to find the one that suits your system. The breathability measure you're looking for is the SD number - you want lower the better. See chapter 6 for more information about paint.

Summary

- There is a major shift going on that includes natural materials and builders are not up to speed. It's for us as homeowners to get what we know is best for our home, family and the climate.
- Natural materials have been used for millennia. They are non-toxic, vapour permeable and contribute to a healthy home.
- There is a wide range of materials from hemp to diathonite, sisal to denim. All suitable in different situations. In particular they will keep your home cool in summer as well as warm in winter.
- You need to use lime or clay plaster over natural materials because they are also vapour permeable. These plasters also help manage moisture in your room by absorbing and desorbing as the air dries.
- Choose an architect/designer that is familiar with natural materials and tell them this is what you want. It will affect the design and drawings they produce. They can also support you if the builders are wary of the unknown.

To find the information indicated by the * follow the QR code.

NOTES

CHAPTER 12
THE CHALLENGE OF SUSTAINABLE HEATING

We've become attached to gas heating. It's efficient and it's been cheap for so long that it's hard to imagine anything else. I recall visiting a friend whose home wasn't on the grid and I found it very strange – how did they manage without gas?

We know now that using gas is a major contributor to climate change, releasing massive amounts of carbon into the atmosphere, so we need to get off it ASAP. But what are the options? It's a minefield right now, with some keen to get on board and some determined they won't. And we're not helped by the media constantly posting bad news stories.

So where do you sit? Would you have a heat pump? Try asking your friends and see what answer you get.

The most common reply I hear is that heat pumps don't work. They're too noisy, won't keep your home warm and cost a fortune to run. This is the standard rant of the media - and many builders - so it's easy to get caught up, with partial understanding, writing off something that has to be part of our future.

My suggestion is to ignore the press and the builders. Instead go speak with someone who has a heat pump. There is even a website set up to do just that. Visit a Heat Pump* gives details of people who own one and will invite you into their homes to see, hear and feel the impact of the pump.

WHAT ARE HEAT PUMPS?

There are two forms of heat pump:

- air source heat pump (ASHP)
- ground source heat pump (GSHP).

The air source heat pump is the most common because it can be installed easily in most locations. It works by pulling heat from the air, using that to heat the home. I know what you're going to say: How does that work in winter when there's no heat in the air?

It is a mystery to me, too, but they've been used in Scandinavia for years and they have extremely cold winters.

It works a bit like a fridge backwards. If your fridge has exposed coils at the back, you'll know that these can get very hot. The heat pump is just using that heat for our homes.

The pump contains a refrigerant that absorbs heat even when the air is freezing outside. As long as the temperature is above absolute zero ($-273.15°C$) there will be heat and energy to capture. Once the refrigerant has absorbed heat it becomes a gas, which gets even hotter when compressed, at which point it flows through a heat exchanger and into your central heating system. As it cools down the refrigerant turns back into a liquid, its pressure drops and it's ready to absorb heat again.

Ground source heat pumps are more efficient than air source because they draw heat from deep down in the earth. A series of pipes are laid in trenches about 2m deep or in vertical bore holes about 100m deep – roughly 150 adult steps. At these depths the ground maintains an average temperature of 10–12° all year round. Once the heat is collected it goes to the heat pump and follows the same pattern as in the air source pump.

I have visited a home that has a ground source system. Internally the gubbins is pretty much the same. The positive difference with ground source is the consistency of the heat. Whereas the ASHP needs to work harder in winter, the GSHP plods on in the same way all through the year, so won't cost more when it's cold.

The downsides are the cost of digging those bore holes or trenches, and the space they need. I understand the big cost is bringing in the machinery to get the job done. You also need the space to accommodate the pipes – either a lot of ground (2.5 times the floor area of your property) or a space to get in machinery to bore down 100m.

ALTERNATIVE FORMS OF HEATING

Heat pumps are the most popular forms of sustainable heating, but not the only ones. You could also have:

- biomass boilers
- geothermal systems
- electric combi boilers.

A biomass boiler forms the basis of a central heating system that uses organic matter for fuel. Different from a wood burning stove, it provides heating and hot water for the whole house, rather than just one room. It burn logs or pellets, which have to be constantly fed into the stove to keep it going.

The biomass boiler is considered to be carbon neutral, because it burns wood that has been grown for the purpose and as soon as it's cut, more trees are planted to absorb carbon. So like for like in effect. Depending on your viewpoint, this may or may not answer your climate concerns.

Geothermal systems access heat from deeper down in the earth than ground source heat pumps. Simple in countries like Iceland that have geothermal energy naturally, less useful elsewhere. There would also be very high up-front costs in order to dig down far enough to find the level of heat needed.

An electric combi boiler is a possibility, especially for smaller properties, since they're efficient, compact and quiet. They produce no direct carbon emissions, so are a more environmentally friendly option, especially when paired with renewable energy sources. The main problem is that they would be expensive to run for a larger house or family.

Open house? Only in the winter

I did my first Open House in summer 2022. So many people were interested in what we'd done to our house – retrofit, heat pump, solar power – that I decided to invite people in to see.

Then I realised – there's nothing to see in an eco home! Everything of interest is hidden behind the plaster. Once completed, it just looks like an ordinary home. Fortunately I'd taken plenty of photos, so got the relevant ones blown up. They're a real help in illustrating what we did. But still we walked around the house talking about the impact of our work rather than experiencing it.

After my first day of running four back-to-back 90-minute sessions, I made three decisions:

- *four sessions in one day is far too much!*
- *this was worth doing. People came in very unsure and left with a sense of possibility*
- *I would only run them in the winter.*

When people walk into the house in the winter, you can hear the sigh as they relax into the warmth. I'm no longer just describing the level of heat in the underfloor heating – they feel it. I don't have to tell them the heat pump is quiet – they listen. And there's no more discussion about space taken up by internal wall insulation – they see for themselves in the window reveals.

End result? A few more people open to the idea of making their home energy efficient. That's good enough for me. That and the fact it makes us tidy up!

The resistance to heat pumps is all about change and a willingness to put our money where our mouth is. Most of us take a while to warm up to something new. It's easier to stick with what we know and claim we're too busy to do the legwork.

In my lifetime change has been incredibly rapid so I know what I'm talking about. We can't be bothering with the demands of change, but we also like a constant flow of new, shiny things to be put in front of us. The end result is that we over consume in some areas and don't want the faff of change in others:

- I was determined to hang onto my Psion 5 digital assistant – it was brilliant. But as I saw more and

more people with an iPhone, I began to appreciate its merits
- I was determined I'd never be caught wearing skinny jeans – until I got my eye in and anything else looked weird
- I never thought this new-fangled internet idea would catch on. I knew I could look stuff up but why would I want to? Now I struggle to live without it.

The turning point, I realise, is when I've seen and heard about the new option enough times. It's as if there is a switch in my brain and until that clicks, I won't change. Once it clicks, I must change. It's what advertising is all about and how social media works as a promotional tool.

It's also about keeping up. We want to be seen as equal to others, so when everyone else has something we've resisted, we'll soon get on board.

We're in that stage with heat pumps

The trick is to make sustainable heating sexy. Then we'll be able to flick the switch and make life safer for personal wellbeing, future generations and the planet.

Ever heard of Everett Rogers? He was a communications professor who introduced the concept of 'early adopters' in his* 1962 book, *Diffusion of Innovations*.*

According to his theory, innovators come up with great new ideas and early adopters are the first to take them up. These are the people who enjoy change and like to try out new ideas and new products. They're followed by the early majority, who decide it's time to get on board – this is the bulk of people. The late adopters come next and it's only the 'laggards' who are left.

In heat pump terms, the early adopters are totally engaged and enjoying their new-found freedom from gas. It's because they are so excited about what they have that a site like Visit a Heat Pump* can exist. They are delighted to share their experience of living with a heat pump in the hope of encouraging others to follow suit.

They want to encourage others for all sorts of reasons:

- they know their house is warm and that the heat pump works
- they know they are saving money on energy bills
- they know we can't continue using gas for the sake of the grandkids.

Now it's time for the early majority to get involved. This will start the snowball effect: the more people talk about heat pumps, the more others won't want to be left behind. And as they too wax lyrical about their heating, heat pumps will become the new normal. And then we'll really be on our way.

WHY DO HEAT PUMPS MATTER?

In fact the usual question is: what's wrong with a gas boiler? They've worked fine for years so why change now?

If you've read chapter 4 then you know exactly why. So forgive a quick recap for those dipping in.

Here's why: gas is a fossil fuel that's been storing carbon safely while trapped in the earth. As soon as it is pulled out of the earth and burned, it produces carbon dioxide, which holds heat in the atmosphere and, as the gas builds up, we move closer to major climate change.

We already see the results of this extra heating. 'Exceptional' weather events become less exceptional each year. Summers are warmer, floods more common, forest fires regular occurrences. As climate changes, food production becomes more problematic. The list goes on.

And, sure, just you having a heat pump won't stop the climate emergency. But remember: we all need to hear about new things a number of times before we engage. So if you have a pump and talk about it to the people around you, you help nudge them towards a heat pump of their own. At which point they start talking about it and momentum begins to build.

Let's look at the objections

Because we're in the middle of a change process, there are sure to be objections. So let's take a look at what they are about:

- heat pumps are expensive to install
- heat pumps cost a lot to run
- they are noisy and neighbours will complain
- I don't have space beside my home to put it
- there's no room inside my home for the gubbins
- I'd need to change all my pipework
- there's not enough insulation in an old house
- I'd have to have new radiators
- it won't keep us warm in winter.

Heat pumps are expensive to install

Of course they are. Everything costs more when fewer people are buying. When I bought my first MacAir laptop I paid over £3,000. Now I can get the latest version for £849 and it's a bigger and better machine in terms of capacity. Because more people are buying, prices come down. It's economies of scale.

To help solve this problem the government is giving a grant of £7,500 to each new heat pump buyer. It's not means tested and your supplier will manage the application for you. Remember this when you get a quote in - some providers tell you the total cost first, then take off the grant, so it can be a bit of a shock!

The grant is available until 2029, but we've no idea what will happen after that. So if you're planning a boiler replacement get it done before then to be on the safe side.

And costs are already coming down - Heat Geek* have come up with a brand new way of planning and installing heat pumps – Zero Disrupt. Now a heat pump can cost you the same – or less – than a gas boiler.

They have developed a system, using AI, where they can survey your home, work out a plan of action with you and give you the cheapest heat pump design with minimal disruption.

In addition Heat Geek will guarantee a SCOP (seasonal coefficient of performance) of 1:3.3. (see below)

Heat pumps cost a lot to run

This is a challenging one to answer. At the time of writing, electricity costs a lot more than gas, despite the fact that the majority of the electricity used is from renewables.

There are three reasons for this:

- **The marginal wholesale price**: The wholesale price of electricity is set by the cost of producing the last unit of electricity needed to meet demand. Which means the grid uses the cheapest energy first, then works its way through to the most expensive at the end of the day – which will be gas. And this is what sets the

price for everything*. This is what people mean when they say the price of electricity is tied to gas. And also why companies like Octopus and Ecotricity are working hard to balance the grid without having to use gas, in the hope of taking the higher cost out of the equation.
- **Carbon taxes:** Electricity is subject to carbon pricing. This means consumers are paying carbon taxes on electricity used at home. Interestingly, there are no carbon taxes applied to gas or oil for residential use.
- **Policy costs:** funds for some social programmes that tackle fuel poverty and promote renewable energy generation have been placed on electricity bills, rather than on gas bills. Although at the moment the Energy Price Guarantee has shifted this onto general taxation rather than levies on bills in order to take some of the pressure off.

So until the cost of electricity is unhooked from the price of gas and taxes are spread more evenly, a heat pump will cost more than it needs to.

But don't turn away yet…

Coefficient of performance: there is a remarkable fact about heat pumps that balances out the excessive cost of electricity. This is the coefficient of performance or COP: the measure of how well the pump is performing. COP is the measure at any given moment; SCOP (seasonal coefficient of performance) is the average performance over time.

Put simply: for every pound you spend on gas, your boiler will deliver £0.70 worth of energy back to you. For every pound you spend on electricity, your heat pump will deliver between £3 and £4 worth of energy back to you.

This is expressed as being a SCOP of 1:3 or 1:4 and it will depend on your level of insulation, how many radiators you have and how good your installer is. You can sometimes get a higher SCOP by putting in larger radiators, improving your insulation and lowering the flow temperature of your heat pump.

This means the until the pricing system changes, the heat pump will cost about the same as a gas boiler to run. But think about when sense prevails and we're paying the actual cost of renewables – then you're really going to be quids in.

They are noisy and neighbours will complain

Heat pumps can be noisy, but it's usually to do with the installation.

Lack of experienced installers is one of the main problems and the actual source of many complaints about heat pumps. I love it when real techies talk about the problems people complain of. They get right down into the nitty gritty of installation and can identify a clamp that's been put on backwards or a pump that's not sitting balanced on its plinth. Truth is, there's nothing wrong with the machines, but how they're put in can sometimes be cag-handed.

So if you want a heat pump, put your effort into finding the right installer. Start by looking on YouTube at the Heat Geek* channel. Adam Chapman, who set up the channel, has become a real expert in heat pumps and electric heating and his videos are easy to understand. Have a browse and you'll learn enough to question any provider thoroughly.

In terms of the general noise of a heat pump, I haven't heard any complaints about excessive noise that can't be put down to installation. Our pump sits on the flat roof above our main living area and we rarely hear it. Only on extremely cold days will we hear a gentle chug. But then I could always hear the gas central heating when it was on.

In the past three years heat pumps have got a lot quieter, but that does mean they've had to get a bit bigger. This is probably the biggest differentiator between expensive and cheap heat pumps. As a rule the more expensive they are, the quieter they are. This is managed by adding more internal insulation, using better components and improving the build quality.

I don't have space beside my home to put the pump

This can be a problem if you've little space outside your home. Some terraced houses, for example, may only have a small area outside, making it difficult to find a place for a heat pump.

- **Distance from neighbours**: The one metre between the heat pump and the boundary of your neighbour's home has now been removed, but there is still a rule that the noise mustn't exceed 37 decibels (dB) at 1m from a neighbour's window (37dB is equivalent to the hum from your fridge).
- **Where to place it:** if you have any garden space then a pump can be placed further away from the house. You can also box it in to a certain extent, but take care not to impede air flow, because that'll make it less efficient.
- **If you live in a flat**, it is possible to have pumps that sit on a balcony. Your alternative is to get together with your neighbours and look into having a pump that'll serve a number of flats.

I recently went to visit 800-year-old Lambeth Palace in London* – a massive old building that is served by three heat pumps. And they are set well away from the building. So if it can be done there, I'm sure there is a way if you're willing to look for it.

There's no room inside my home

The heat pump does need space inside the home. As a rule of thumb if you have a hot water tank you'll be able to house the internal workings of a heat pump. The water tank and header tank can go anywhere in a house although, as with putting it further into the garden, you'll pay a bit more for the extra piping and work to install it.

If you don't have that space, there is now the option to have a very small water tank that works more like a combi boiler. This has been developed by Heat Geek to fit into a kitchen cupboard, on top of a cupboard or be hung on the wall like a combi*.

The other option is the Sunamp* heat battery. You install this in place of a hot water cylinder. It's smaller than a standard water cylinder, it's tried and tested and generally works well, so worth exploring.

I'd need to change my pipes

You may need to change pipework in your home in order to install a heat pump, but it's not a given. Reasons why you might need to change include:

- you have microbore pipes and they don't suit the heat pump you want
- someone has installed plastic pipework in the floor
- you don't have enough insulation and your house is old.

Microbore pipes: I always assumed microbore pipes were in old houses, but it turns out they were installed from 1990 onwards because they were cheaper and easier to install. As the name suggests they are of smaller diameter – less than 15mm.

If you want to check what pipes you have:

- microbore pipes are 10–12mm in diameter. About the size of your little finger
- regular pipes are 15mm or larger. About the size of your middle finger.

The reason microbore can be a problem is the flow rate required for a heat pump. It may be higher than the pipes can manage, which can cause blockages.

However, changes are not automatic. Heat pumps are developing at a great rate, so you may find you get a very modern one that can manage the smaller pipes. Also if your home is well insulated, the requirement from the heat pump will be less so the pipes may not be an issue.

Plastic pipework: plastic pipework can be a problem and may need to be removed. It all depends on the thickness of the pipe walls. Because plastic is thicker than metal, the useful internal diameter of the pipe is reduced, restricting the flow of water. And that may not suit the heat pump.

Not enough insulation in an old house: this question always amuses me: do I need to improve my insulation if I want a heat pump?

The answer is yes, of course. This is also the answer if you ask the same question in relation to a gas boiler. Any form of heating will have to do less work with improved insulation. If you're in a poorly insulated house, you'll have to run any heating system to full capacity to stay warm.

We've become lulled into a false sense of security because gas boilers can manage sudden shifts in demand – like when the wind is in just the right direction to target your draughty floors or when the snow suddenly comes down. As a result we stop noticing the inefficiencies in the house. This allows us to pretend that the house is OK. When actually it's costing far more than it needs to and increasing our carbon footprint.

It is always worth sorting out your insulation. Whatever you are planning in your house, add insulation and airtightness. Clearly there is a cost to doing this, but there is also a cost of not doing it and the savings will far outpace the costs over time.

Insulation also provides greater mass for the heat pump to work with. The pump begins by heating the air, just like the gas boiler. But because it doesn't switch on and off, it goes on to heat up the fabric of the house. If your walls are just solid brick, it won't be able to hold so much heat. But imagine if you have hemp insulation or wood fibre that has a long decrement delay - the house will hold onto the heat and be toasty warm.

I'd have to have new radiators: let's think this through: your heat pump will provide a flow temperature of approx. 40° so you need a big surface area of heat source to provide the warmth you want. The ideal heat provision for this is underfloor heating (UFH), because this covers a very wide area with a gentle background heat.

If you can't have/don't want UFH, then you're reliant on your radiators and until recently that meant they too needed to provide a big surface area. In fact, if your insulation and airtightness are effective, you may well be fine with the radiators that you have.

You'll need big radiators

We mithered for a long time about radiators. It was recommended that we needed to increase the size over those that we already had. It upset me particularly for the bedroom, where we had a really attractive steel one that I loved. But I also hate being cold!

We had decided on a wooden floor throughout the kitchen and dining room. Since all that was being lifted up so we could insulate under the floorboards, it made sense to include UFH while we were at it. I have never regretted that decision – it's brilliant. Not too hot so you get swollen feet, just a very gentle heat that you really only notice when you step onto floor that doesn't have it.

But then we had to decide about the carpeted front room and the bedroom. I remember the day we went radiator shopping. I'm afraid to say I was being very sulky. We had a choice of one – and it was industrial. But what can you do? If we didn't take the advice we might have had to lift the floors at some point in the future and that wasn't an option.

With hindsight and the knowledge I have now, I'd have got a second opinion. Maybe someone else would have understood more about what we were doing in terms of retrofit.

It is a feature of retrofit and renovation that small items blow up into major events when the pressure gets too much. These radiators were just such an event. In reality we hardly notice them now – they just blend into the background.

Be ready for this: when the heat pump is first commissioned (switched on) it will take the house up to three days to warm up. So don't panic when you feel cold on day two. You'll also think the radiators aren't working if you feel them or that the heat pump is switched off. It's astonishing how the room warms up, even though the floor or radiators don't feel particularly warm. Same with the **UFH**. It's easy to forget it's on, because the heat is so low.

Also don't worry if the newly switched-on heat pump is a bit noisy. It's working hard to heat the home up from scratch. You'll only hear that level of sound again if there's heavy snow on the ground.

It won't keep us warm in winter: this is the standard cry from people who don't know heat pumps. I can safely say I've never been warmer in our old Victorian home in all the 45 years we've lived here.

So why don't people trust that? Essentially, we're confused by how they work:

- a gas boiler runs at between 60 and 70° while a heat pump runs between 40 and 55°
- a gas boiler comes on and off when we need it. A heat pump stays on
- no thermostats are needed for the ASHP – in fact they are detrimental.

All of which spooks people – because we just don't understand the principles.

This brings me to one of the real problems with a heat pump. The problem that sits alongside the need for more experienced installers. And that is us – the homeowners.

WE NEED A CHANGE OF MINDSET

Heat pumps work differently, as outlined above, and we don't like it. With our gas thinking minds, pumps make no sense at all, so we just carry on the way we're used to.

You could do better with that heat pump!

We were very happy with the output from our air source heat pump – 75% reduction in energy usage was a lot more than we expected. We'd hoped for 50%, so this was a dream come true.

Then the Everything Electric Show YouTube channel did a film about the house and what we'd done to future-proof*. I spoke proudly of the fact that we heated the house on the underfloor heating in the kitchen, plus two towel rails upstairs. Even mentioned the fact that we'd taken out radiators that we didn't need.

Imagine my surprise when comments came in suggesting that we could be even more efficient if we changed the way we ran the heat pump. I found this very hard to believe, but there were so many comments in the same vein over the first few days that I had to look into it.

Much to my surprise I realised that we were running the ASHP as if it was a gas boiler – which is completely the wrong thing to do. Running a heat pump efficiently is counterintuitive:

- *when we're cold, we turn the flow temperature of the pump down*
- *to save money, we need to turn all the radiators on.*

Crazy making! So I clearly had a lot to learn. Thank heaven for Everything Electric and the introduction to their followers.

When we live with gas central heating we expect the house to be warm when we get up in the morning and when we come home at night. Those working at home will set different parameters on their thermostat system, but still expect periods of warm and cooler temperatures.

Gas boilers are quite capable of doing this because the higher flow temperature means they can warm the space up quickly. The boiler is on for a short time, heating up the air in the required rooms. Once the thermostat goes off, the heat quickly dissipates.

Here's where the gas mindset becomes a problem. If we run the heat pump in the same way, we will be cold. This is because the heat pump works best with a low flow temperature: around 40° but certainly no higher than 55°. At that rate it's really hard to warm a room up quickly and for a short time. It will just be getting going when it's time to stop.

So what happens? We stick with what we know and blame the heat pump because it's not delivering as we expect. No one talks about this much, so we just carry on doing what we've always done, hoping for a different outcome.

Think of your fridge. You don't come home from work in the summer thinking: 'I need my beer to be cold – I'll pop the fridge on'! You just leave it set at 3° and it switches on and off in order to maintain that temperature. The heat pump is the same – tell it you want the house to be 21° and leave it to do its thing.

It's time to change our thinking

We were definitely gas thinking – albeit still getting a great reduction in energy usage. I've looked back to see how this had happened. I can't recall anyone speaking to us about it, so either:

- the provider didn't understand the way we needed to change our habits either
- he told us and we just didn't take it in.

I hope it's the latter. And it would make sense. Anyone who's gone through a retrofit or major renovation will know just how intense the communication can be. You face a barrage of questions, need to give direction, hold tough conversations to make sure you get what you want… It goes on and on. So one more set of instructions may well get lost in the melee.

ASHP THINKING

Get hold of this and you'll be really happy with your heat pump:

The gas boiler needs you to be warm for a short time so it sprints to get the house ready for you. It works fine until you have to get up in the night, come back early or decide to work from home. Then you'll struggle to stay warm.

The heat pump give you a warm house all the time. It doesn't matter when you'll be in or out – it'll always be ready for you. You set the temperature you want and leave it to deliver. No need to switch your pump on and off. It heats the air, then goes on to heat the fabric of the house, which will in turn radiate heat back into the air.

Heat pumps are at their most efficient when they run constantly. It's the switching on and off – called cycling – that uses up energy, so your aim is to have it chug on like a marathon runner. For this reason it's very important to get the appropriate size of pump for your home. If it's too big it will cost you in constant cycling; too small and it won't keep up.

The size of pump is determined via a heat loss survey that includes:

- property size
- insulation
- windows
- local climate.

Together these determine what your heating needs will be.

One caveat: if you are working on your home, the heat pump installer will need to know your plans with regard to insulation and airtightness, as well as the U values you're working to. From that it should be possible to work out the right size of heat pump for you.

It's worth getting a couple of recommendations for the size of pump you need. If your provider is not very experienced they may opt for something a bit bigger 'just in case'. But in heat pump terms that's going to cost you in efficiency.

Why don't I need thermostats?

The very idea of not needing thermostats on a heating system hits at all our gas boiler conditioning. 'Of course we need a thermostat – how else will we manage? There will be no control at all so the pump will never switch off!'

Weather compensation: heat pumps are regulated by weather compensation. This is a control on the heat pump that automatically adjusts the output temperature based on the outside temperature. It has an external sensor that monitors the outside temperature and changes the flow temperature of the pump to suit. When the weather gets colder the system increases the water temperature; when it gets warmer it reduces the water temperature to deliver comfort with greatest efficiency.

You can use a thermostat and weather compensation, but you run the risk of too many zones of heating switching on and off, which means there won't be much water flowing round the system. Heat pumps need a specific flow of water to work, so if you do have a lot of zones you're more likely to need an extra tank called a volumiser (sometimes known as a buffer) to make sure there is always enough water running round the system. This is slightly less efficient, adds cost and takes up space. So the best way forward for greatest efficiency is to cut out the thermostats.

It's always worth checking that your installer has turned on the weather compensation. It's not always the case and it will cost you money in terms of efficiency if it's not on.

Can heat pumps also cool my house?

Heat pumps can be set up to reverse the process – cooling your house on hot summer days – if the controls are enabled. This isn't a straightforward process, so go exploring and ask a lot of questions before making your decision.

You need to remember that cool air sinks and hot air rises. If you live in a place where the weather gets really hot, then you need cool air to go in near the ceiling so it can cool the room as it drops down. If you live in a place where cold weather is the biggest issue, you need warm air on the floor or low down so it warms the room as it rises.

If excessive heat is a big problem, then an ASHP may not be the way forward. You'd be better off insulating your home with natural materials that can manage the heat for you. But in a cold country with occasional heatwaves, having cool water flow through your underfloor heating pipes might be appealing.

Summary

- You have different options for sustainable heating, the most popular being the air source heat pump.
- Changing from gas central heating is a major shift — and is important because of the urgent need to reduce carbon emissions.
- Despite the numerous objections to heat pumps, the main problems are lack of experienced installers and customer mindset. Using a heat pump like a gas boiler costs us money.
- You will benefit from improved insulation whether you have a gas boiler or a heat pump. Running a home without this will always be more expensive.
- Look into options for a cooling setting on your heat pump to help you manage excessive heat in summer.

To find the information indicated by the * follow the QR code.

NOTES

CHAPTER 13
HOW TO GET THE RENEWABLE ENERGY YOU WANT

I f you want to be energy efficient, save money and reduce your carbon footprint, then solar is the obvious next step. Why pay out to heat your home or drive your car when you could let the sun do it?

It's unlikely a builder will suggest this – it isn't part of their specific brief – unless they're truly eco builders. So if your panels aren't already on order, let me encourage you.

You'd think that solar would be automatic by now. The price of panels is going down and it's a very short installation process. Once your panels are ordered you'll be up and running in a couple of days.

But still there are endless empty roofs out there that could be generating energy for home and/or the grid. I'm not sure what it's going to take to get the message across. At least soon it will be obligatory for new houses to be built with solar already installed.

Chatting with the accountant

Not sure about you, but I find it hard to focus on discussions with the accountant. I end up thinking about mopping the kitchen floor, sorting the airing cupboard or doing the shopping. Anything other than money and numbers!

But in one such conversation, my ears pricked up. He was talking about our desire to have solar energy. We wanted solar but weren't sure we'd have enough money to do it for a while.

He asked about our energy consumption, what it cost us and the investment it would take to install solar panels. After some quick calculations he made a proclamation: "You'd be better off using some of your savings to buy panels than hoping for a better interest payout."

That was the best meeting I've had. And great to hear that solar was such a good investment. You can guess where my next call went to...

There are a number of aspects to solar for you to consider:

- which direction your roof faces
- whether you need planning permission
- levels of shade
- whether to include a battery
- what type of panel you want
- what happens during installation
- formalities you have to go through
- how much you could generate
- whether you could sell back to the grid and what tariffs are available.

Direction of your roof

The biggest question when it comes to having solar is the direction your roof faces. The more direct sunlight you have on your panels, the more electricity you generate.

In the northern hemisphere:

- a south-facing roof is your best option. It will provide you with the most direct sunlight
- east will give you sunshine in the morning
- a west-facing roof gives the option of sunshine in the afternoon and evening.

In most circumstances there's little point in putting panels on a north-facing roof, because they only produce useful amounts of electricity in the summer, when you don't need much. And they are more prone to lichen and moss growth. If this is your only option it may still be worth exploring, if you have money available and have a lot of battery storage to soak up that summer production.

Our house faces east–west, so we have five panels on the front roof and three on the back. This clearly won't give us the very best of sunshine, but it provides enough to make a good dent in our energy bills. Given that we didn't get a battery for them, we were often donating energy back to the grid. So, even with east–west panels, we made more than we could use.

Your house roof may not be your only option. If you have garden sheds or garden rooms, check out the direction of the roof. Remember to check the solidity of the roof, too – it needs to be strong enough to support the panels.

The same applies if you're building an extension – bear in mind that this could take solar if you position it correctly or make roof space available. Be clear what you want as you start the planning process – don't expect a builder or architect to suggest it. Sustainability and energy saving aren't yet hard-wired into the construction psyche.

The other time to think solar is when planning a loft extension - where you put a window will impact your ability to use the roof for panels. We have a Velux window in our loft. I'd have thought twice if I'd known. If the window wasn't there, we could have two extra panels! So add solar from the outset with any renovation/retrofit planning you're going to do.

To get the very best out of panels, they should be at a tilt of 30–40°. This is optimal, so if you can't get to that exactly it's still worth considering.

It's also possible to put panels on a flat roof as long as there is space to tilt them towards the sun. These are normally kept at 15° on frames that use ballast to hold them onto the roof. Remember that you'll need space between the panel rows so they don't shade each other.

Do I need planning permission?

The quick answer is probably no – unless you're in a listed building or part of a World Heritage Site. If you're in a conservation area check with your local authority first. Government guidelines in the UK have made this easier, so you may not have a problem. Otherwise solar panels come under permitted development so you can go ahead whenever you're ready. We only had to make sure the panels didn't sit out a long way from the roof, which would actually be quite difficult to do anyway!

Levels of shade

There is more to consider that just the direction of your roof. Do you have loads of trees around that will throw shade on the panels? If so, it will be recommended that you take the trees down. It's up to you whether you actually do that.

We have some shading from trees but they are the sort that lose their leaves in the winter, so it's only at high summer that the panels are in shade. We have yet to decide what we're going to do. It's a difficult balancing act – cut down trees and release carbon in order to increase energy production? Or use what we can get, since it's only summer that's affected, and leave the trees alone?

Check out any buildings that might impinge on the panels. Maybe your neighbour blocks the sun at certain times of day or maybe they have trees that you're not in control of. Again it is your decision whether to have the panels.

The downside to going ahead even if conditions are not optimal is that providers are unlikely to give a guarantee about how much energy you can expect to produce. Which is fair enough – if you don't follow their advice, you won't get their confirmation. But it's still your decision. You may decide you want some solar energy even if you can't get the optimal amount. Or you might only countenance it if you can get full value. Both are valid.

Microinverters: if shade is an issue for you then you need to know about microinverters.

There are two main ways of wiring up the panels:

- string inverters
- microinverters/optimisers.

With the **string inverter** one panel is wired to the next and the next and so on. Hence 'string'. It's been the standard way of setting up photovoltaic (PV), but it means that if one panel has a problem or is in shade, all the panels are affected – one panel in the shade or damaged, producing less or no power, means all your panels are knocked out.

Microinverters/optimisers are the solution to this. Each panel has its own microinverter, so if one panel is in shade that's the only one affected. And if by any chance, one panel should go off or have a problem, the others can carry on doing their job. So this is definitely one to include in your requirements when talking with providers.

Build a solar roof

Our house is east–west facing, so when we fitted our eight panels shortly after the retrofit, we didn't expect a huge output. I'm pleased to say we were wrong – they do a remarkably good job and we were feeding the grid way more often than expected. But we wanted more. The idea of being self-sufficient – for some of the year, anyway – was very appealing.

We considered putting more panels onto the garden office. It would have meant reinforcing the roof and it still was only east-facing. So John came up with the idea of building a solar roof. If we took down the existing office and the small pergola that held our blow-up hot tub, we'd have a space that could be entirely south-facing.

That was an exciting proposition – a solar roof! And not only that, we could put all our new-found knowledge into a sustainable build – without bricks and concrete. For a couple of old building bores that was very exciting!

We found an architect in the midst of her Passive House training and she was as excited as we were.

Plans were produced with an emphasis on the largest roof possible.

Then came long discussions with solar providers, with the main question being 'how many panels can you put on this roof?' The answers were varied – ranging from six to 10. We learned about the different options from aesthetically pleasing integrated panels to the more utilitarian options. We discovered that the least lovely were the most efficient, so utilitarian it was.

We managed to find a supplier who joined in the adventure, making suggestions for how to adjust the plan of the barn slightly to give greatest value before announcing that we could have 10 panels.

So now we have 18 panels and a battery. And it's great. The summer months will pay for the full year's power and we make some cash as well. Game on!

Whether to include a battery

This is a thorny question, especially if you're concerned about the environment. There are a number of issues:

- if you want to sell energy back to the grid you need a battery
- if you want to access cheap energy at night, you need a battery
- a battery will help you balance sunny days so you can use your own energy in the evening.

Having a battery definitely helps you make the best of the energy you're producing. On sunny days, you'll end up with plenty to run the house for the evening. Once winter closes in, sunshine is less and the days are shorter, so your battery takes on a different role. Now you can stock up on the cheap energy from the grid during the night.

This is a benefit for your bank balance and the climate. Electricity produced overnight is actually some of the greenest energy available. Energy companies are currently paying wind farms to switch off overnight because they can't use the power. So by having a battery you help balance out the grid, storing this green power ready for when you need it. Not to mention that you can pay as much as 40% less for nighttime energy!

Back-up power: the possibility of back-up power to cover you in a power cut is an interesting one. I assumed this would be automatic, but I was wrong. If you want this capability, tell your installer when you first meet – including how much power you want to have available. The options range from powering a couple of sockets when the main power goes down to backing up your whole house.

- The battery must be able to disconnect from the grid. When there's a power outage everything is switched off to allow people to get in and mend it.
- If your battery starts releasing power without disconnecting your house from the grid first, some poor unsuspecting soul will be hit with a bolt of electricity
- To avoid this, for a split second the power will go out, then it will come back on again, feeding directly from your battery with no reference to the main grid. When the mains power returns, the battery will do the same again and connect you back up to the grid.

Environmental impact of a battery: this is important to think through because it is a problem.

The manufacture of solar panels needs silicon, silver, copper, aluminium, indium. Solar batteries need: lithium, cobalt , graphite, nickel, manganese. All these minerals need to be mined.

There is of course the environmental damage of mining, but also the wellbeing and welfare of those people who actually do the mining. It's known that company owners don't always stick to great standards of care*.

Increasingly there are moves to recycle the minerals used in batteries (car batteries, too) and this will only increase. However, it is a decision for you to make when considering solar.

We need to balance the benefits of renewable energy and the environmental cost. The carbon usage of batteries and panels is front-loaded – it mostly takes place during manufacture. Once that carbon load is paid off, then they become carbon neutral.

I struggle with this one and we didn't have a battery for a long time because of it. I realised a long time ago that just by living on the earth we have an effect. As Jane Goodall* said: 'You cannot get through a single day without having an impact on the world around you. What you do makes a difference, and you have to decide what kind of difference you want to make.'

Look for the least worst: it's impossible to live without causing some harm, but I can do my best to reduce the damage. I applied that to the question of solar and decided that the more energy we could produce – sharing with the grid when we could – the less gas would need to be used to balance up the demand.

You must work it out in your own way. Whatever you do must feel right for you.

Bi-directional charging: just before we move on from batteries, a quick word about bi-directional charging. If you have an electric vehicle (EV) you may be able to use that battery to power your house. This is new technology and it saves you buying an additional battery.

I still find it hard to take in that a car battery can drive me where I want to go <u>and</u> power the house. I have visions of not being able to go out because I've put the washing machine on. But I am assured the car battery is much bigger than needed for the car and the relative use of energy in the house significantly smaller.

If you are planning an EV then add this to your list of requirements. Only a few car models and manufacturers offer this at present and you'll need a special bi-directional charger fitted so you can take advantage of it. But it is certainly worth consideration.

WHAT PANELS DO I NEED?

Solar energy is very fast moving, so if you're starting this part of your retrofit journey make sure to stay fully up to date with new developments.

What happens during installation

In my experience it's remarkably painless and I've heard others say the same. A lot of the work goes on outside the house, with just the inverter needing to be installed and wired up inside. For our first eight panels on the house it was just two days for the work to be completed with the panels up and running.

You'll find that scaffolding is needed – not just for the installers to get onto the roof but to make it possible to get the panels up there. They are pretty big beasts! A lot of installers will bring their own scaffold tower with them. This is a question to ask when considering a quote – do I have to pay for scaffolding or is that included?

Other than that, it's pretty straightforward. One of those jobs that's much easier than expected and instantly ready to go.

Measure of efficiency

You are looking for a percentage figure, detailing how much sunlight hitting the panel is converted into useable electricity. The range in the UK is usually between 15% and 25% for residential panels – and clearly we're looking for the higher the better. It's also worth discussing commercial options with your provider. It may be possible to fit panels on your roof that have a bigger capacity. A lot will depend on the size and available space, so make sure they take time to consider what's possible. Don't let them fob you off with the panels they happen to have in stock.

"We can get bigger panels on there"

Velux and dormer windows can really mess up a roof when you're considering solar panels.

We have our loft as a useable room. We didn't have space for a fixed stair so it can't be a bedroom, but it's had many different lives over the years – office space, play room, teenage hangout – and now just the standard loft/treasure trove for my little grandson.

All of which meant we needed a window, so a large Velux sits dead centre.

When we reached the solar stage of our retrofit we were a bit fed up with ourselves. Given the loft is now just a storage space, we don't need a window in the same way and the space could have housed another panel. Fortunately our provider was good at thinking through obstacles.

Solution – he found larger panels that produced more than the residential options. Their size fitted well so they sit around the window on two sides with no wasted space. The third side is hampered by the chimney, so it does look a bit odd, but they're producing like a good 'un so who cares!

We were fortunate with this. Some of the larger panels are just too big to be put safely on the roof, having been designed for solar farms, but it's always worth an ask.

TYPES OF SOLAR PANEL

There are two types of panel:

- thermal solar – these focus purely on heating your water
- photovoltaic (PV) – these are what we think of as solar panels and they convert sunlight into energy.

Solar thermal panels (solar collectors) absorb the sun's heat and use it to heat up water stored in a cylinder. They're more efficient than PV panels because heat waves carry more energy than sunlight, plus there is no process of converting into electricity, which takes power in itself.

They work in cold climates and in overcast weather, and have their own energy storage system. It's a great option if you don't have enough roof space for a full PV array. The big advantage is that it will heat all your hot water in summer so you can switch off your normal heating system altogether.

While solar thermal only carries a five to 10-year warranty, they're known to last for a lot longer than that – up to 25 years. Once they're in place you don't have to do anything – they just plod along on their own. No servicing needed, so once you have the installation cost that's it.

Solar photovoltaic (PV): the world of solar is opening out all the time, with ever more efficient and attractive panels becoming available. If you are thinking of solar, then it's worth exploring new initiatives to see what's coming onto the market. At present the primary options are:

- rigid solar panels
- thin film.

Rigid solar panels: if you're like me and walk around solar spotting, then you're probably looking at rigid solar panels. They're the most efficient and easiest to install. So right now they're the most commonly used option.

Once up on your roof, there is nothing else to do. No servicing is required – they just get on with their job. Manufacturers give a 25-year warranty and they'll go on for a lot longer than that. I heard recently about homeowners who are still using the panels they had installed in the late 1970s!

Rigid panels are generally monocrystalline, which are the most efficient, ranging from 15% to 24%. I was surprised the number was so low – I assumed it would be much higher. I went searching and discovered that this is very efficient. Importantly, at 24% they reduce our energy costs significantly and help save the planet. And research is going on all the time, so great strides are going to be made in the future.

They perform well in low light conditions and cloud, so you don't need masses of direct sunlight to create electricity. On the downside, they are sensitive to shading, so you need to think about your surrounding area.

Bifacial PV panels: bifacial panels are a type of rigid panel. There is no backing sheet to the solar cell, so light can be accessed from the front and back. We have these on our garden barn. I didn't know they existed until they arrived in the garden. Turns out they were the most up-to-date panels at the time with the highest efficiency.

Mind the birds: one last comment before we move on from standard rigid panels. Because they sit up slightly from the roof, there is a gap between the panel and the tiles. Perfect place for birds to nest!

Ask about this when you're vetting providers. You may be told that it won't be a problem – and of course it may not be. But if it does happen, you're left with a constant rain of bird droppings down your wall or onto your windows. There will be all sorts of mess and noise as a result. And it would be much worse to deal with the situation once it's happened. Better to just stop the birds nesting there in the first place.

This wasn't done on ours and we weren't told about the possibility. Then we noticed that a lot of pigeons were mooching around looking for their next nesting place. End result: we had to pay the provider to come back and install netting for us – an extra cost that could easily have been avoided if this was done while the scaffolding was up there.

Roof-integrated PV

The clue is in the title: roof-integrated PV panels are built into the roof – they look fabulous! It's the perfect option for a new build – they look really smart and mean the house is instantly more cost-efficient. Plus you don't have the cost of roof tiles and no gaps for the wind or birds to get under.

The panels sit flush with the roof, so the whole thing looks streamlined and elegant. When installed on a new build it reduces the cost of tiles. They are less of an issue for any planning permission because they are so unobtrusive and they're light so don't add significant weight to the structure.

Integrated panels would be perfect for specific conservation areas where PV is not part of permitted development. However, installation then would be a much bigger issue since it would require a reworking of the roof. But still, if you are in such an area it might be worth looking into as an option.

Solar tiles

This is a fascinating advance in PV provision. Instead of your normal roof tile, you can have each tile as its own solar panel. They are perfect if the look of your home is really important.

There are downsides:

- more expensive than standard panels, although that might be offset by not needing roof tiles
- less efficient. Only 10% to 20% efficiency versus 18% to 24% with a panel. Although you will be covering a much greater area so you may end up with similar amounts
- they take a long time to install.

However, they're more likely to get planning permission on a listed or conservation area building.

If you need a new roof or you're building from scratch, then it's worth a look. But do ask plenty of questions about efficiency and what you could expect to generate.

. . .

Thin film panels

You'll have seen these around the place without taking them in. My daughter gave me a little light to sit in the garden that's powered by solar – that will be thin film. You may even have similar lights in your own garden.

This is exciting stuff. A thin panel will be much less efficient and have a shorter life span, but they open up so many possibilities:

- going camping you can generate enough to power your phone
- small lights for the garden and garden shed
- solar on your car – imagine, when this improves, you could charge your car as you drive!
- a thin film on south-facing windows
- in remote locations – disaster zone relief centres, remote schools, homes that are off-grid
- wearables – imagine having solar cells on the back of your gloves or your hat!

FORMALITIES

Inevitably there are formalities you need to adhere to. You can't just add extra power into a massive, complex system like the grid without some regulation.

Permissions

You have to get permission to have solar panels. This will be managed for you by the provider, but can take a bit of time to come through and you can't begin installation without it.

The distribution network operator (DNO) is the company that owns and maintains the local electricity network, including cables and substations, to deliver power to homes and businesses. DNOs are responsible for connecting new electricity supplies, and are regulated by Ofgem.

You need permission from the DNO if your install is going to be over a certain size, but realistically few people install less than the limit so it's likely you will need permission. I remember finding this very strange at the time – why would anyone want to stop us putting in panels to help with electricity generation? Had I understood better, it would have made more sense.

The DNO owns and is responsible for the local electricity network infrastructure, and by adding solar you become part of that infrastructure. They give permission for your installation, having taken into account how it will work alongside other energy usage in your area. Similar to the issue of using your solar for back-up power, they need to be sure you won't create a sudden surge that could affect the functioning of the grid.

In addition, if you're going to include a battery in order to export back to the grid, then that electricity needs to be incorporated in a way that won't cause voltage fluctuation or overload the system.

If you go ahead without this permission you won't get insurance, it will affect the future sale of your property and you won't get any payment for the power you give to the grid.

Selling back to the grid

You can sell excess energy back to the grid as long as your panels have DNO approval .

How much you can make depends entirely on how many panels you have, the direction they face and where you live in the world.

I am in the UK and we have 18 panels. Eight are facing east/west; 10 are facing due south.

In the first year with all 18:

- we generated 6,448kWh over the year
- our usage over the same period for the house and car was 5,950kWh.

So we were energy self-sufficient for living and driving.

How much comes back on the excess depends on the tariff you choose. I'm not sure we have this right yet, but it's good to know we're not costing anything in energy terms.

To find out what you would make, go to your provider – I'm sure they'll have loads of information on their website.

Intelligent flux

This isn't an advert for Octopus – that's just the provider we're with right now. I'm sure other providers do equivalent packages so it pays to look around.

The intelligent flux package we have means the company takes from our battery at the times when demand from the grid is highest – between 4pm and 7pm. Then it fills the battery up again overnight so we have plenty of power next morning. The algorithm works out how much we need, so how much they can safely buy from us.

This is not only good for the bank balance, it's helpful to the grid: 4–7pm is demanding because it's teatime. Everyone is cooking, bathing the kids, switching on all the appliances. This is the point when energy companies are most likely to resort to using gas to fill demand.

But if we all have batteries, soaking up excess from our solar, this can be used by the energy companies at the busy times and our batteries topped up for the next day when energy is cheapest and greenest. So we can help balance the grid and help the country get off gas.

There are all sorts of agreements you can make. You can even sell your excess energy back to another provider without becoming their customer. You just may be paid a little less per kWh.

GRANTS

There are less grants now than there were, but there are still some options out there for making solar more affordable.

ECO4

If you are on benefits of any sort, look at the ECO4 scheme and see what you're entitled to. It's not just for solar – you may also be eligible for insulation.

Council schemes

It's also worth checking your local council to see if they can help. Some enlightened councils make large-scale agreements with solar providers so residents can benefit from economies of scale. That can make the process cheaper and give you some security.

PROVIDERS

Look around for local providers, as well as large-scale energy companies, since they will all offer package deals at different times. By handling large numbers, they will have economies of scale, which could reduce your bill.

However, please do be wary of cold calls that promise you the earth. They'll have looked at your property on Google Earth so they claim they know what you can have. This happens to us on a regular basis. The Google Earth shots were taken before we had our solar, so they keep telling me we can have large numbers of panels. They also don't look carefully enough so they think my house is twice the size – including my neighbour's roof in the offer!

If you've had a good read here, then been exploring so you're well up to date, even a cold call might be worth a conversation. When a rep comes to visit, be aware they may tell you all sorts about how many panels you can have. This is where your own knowledge will be of value – listen with caution then check against other quotes.

When you get it right, you're going to love it. Nothing better than charging the car or putting the washing machine on when the sun comes out and knowing it's all happening for free!

Summary

- Solar is an important step in energy-efficient retrofit – the perfect way to power your new warm home, making it even cheaper to run.
- Make sure your roof faces a direction that will give sufficient sun to make installation worthwhile.
- Stay up to date with the latest developments – solar is improving all the time and panels are increasingly efficient.
- Solar thermal will heat your water; PV panels will power your home. You can have standard panels, integrated panels or solar tiles.

- Your provider will apply to the DNO on your behalf to get permission to add your solar to the national grid.

To find the information indicated by the * follow the QR code.

NOTES

CHAPTER 14
WHAT YOU NEED TO UNDERSTAND ABOUT GLAZING

The dread of those double-glazing sales calls used to be very familiar, but nowadays we have to go looking when we need new windows. But that doesn't mean you can rely on getting all the information you need or all the answers to your questions. It's still better to be informed before you venture out.

Glazing is an important part of retrofit

When you first get new windows you won't know yourself. There is nothing worse than old, leaky windows for making a room draughty, cold and uncomfortable.

Double glazing: Most people opt for double glazing as a way of holding heat in the house. The window will have two panes of glass creating the insulated unit with argon or krypton gas in the gap. The idea is that the cold air hits the outer glass and can't jump the gas to make the inner glass cold. So the window won't cool the air in the room or invite moist air to condense.

Triple glazing: it's a slow build, but more people are now installing triple glazing. It's been common in Scandinavian countries for years – they've been forced to protect against the cold. Don't make your decision until you've at least looking into triple as an option. They will have three layers of glass rather than two and are much more efficient at holding heat in and cold out.

I can't face Yvonne

My friends in Sweden – Yvonne and Jerry – have a daughter who lives in London. We met for lunch when they were helping her find a flat and they were amazed:

"How do you Brits live? The windows in your houses are AWFUL! Cold, draughty, rattling – it's really bleak!"

When it came time for us to start the window search, I told John we had to at least look at triple glazing or I'd never be able to face Yvonne again. I also knew that triple would be better for our dream of an eco house – and with climate at the top of our touchstone (chapter 1) we had to make it an option.

So we went to a showroom. We looked, we listened and we were convinced. Having lived with them for four years, I'm totally with Yvonne – how did we manage without them?

WHAT AM I LOOKING FOR?

Best to get your ducks in a row before starting your search. To do that you need to understand:

- How window efficiency is measured
- The different types of glass you could have
- The different finishes to choose from.

How window efficiency is measured

When entering into a discussion about new windows the measure you need to listen out for is U value (see chapter 6). This measures how the window transmits heat. Heat going out through your window is the last thing you want, so you need the windows to do a really bad job of it.

When it comes to U values, always go low. You want to lose as little heat as possible. As a rule, values are as follows:

- single glazing = $5.6 W/m^2 k$
- old double glazing = $2.8 W/m^2 k$
- new double glazing = $1.4 W/m^2 k$
- standard triple glazing = $1.2 W/m^2 k$
- high-performance triple glazing = $0.8-0.6\ W/m^2 k$.

Seeing them altogether like this helps to understand the significance. We all know how draughty an old single-glazed window is, so you can just imagine what a U value of $0.6 W/m^2 k$ would feel like. Bliss on a winter's day!

- **Single glazing:** very cold indeed, prone to condensation, leading to mould. A big part of a leaky home and worth taking action on as soon as possible.
- **Old double glazing:** expected to last between 25 and 30 years so if yours are old, check for draughts and cold. (See chapter 10 for how to do this.)
- **New double glazing:** will be much more efficient and better looking, depending on the finish you choose.
- **High performance triple glazing:** more efficient again than new double glazing. Definitely worth considering and not as expensive as you might think.
- **Secondary glazing:** a separate layer of glass added over the top of the basic window. Useful in

conservation areas or heritage builds that don't allow window replacement.
- **Window repair:** companies that renovate and repair. They will remove the glass and replace it with a thin double-glazed unit. This is called a heritage pane. The efficiency won't match a new window but it'll be a great deal better than sticking with a single glaze.

Different types of glass

Retrofit is a new kid on the block and the pace of change is amazing. Glass is just one example of that, so keep an eye on what's new and interesting. Write down any questions before you go – there's a lot to take in so you'll forget otherwise – then you'll be able to find out what's new.

C.U.In glass, for example, is thinner than standard glass and has a film inside the glazing that makes it very thermally efficient. It claims to work in a similar way to triple glazing, with the film acting as the third pane*.

This comes into its own when there is a problem with weight – e.g. in bifold doors where the weight of triple glazing can be hard to manoeuvre. And of course, because it is thin and light, less glass is used, which means a lower carbon footprint.

Different finishes to choose from

It's not always that professionals forget to tell us what we need to hear. Sometimes they tell us everything, but we just don't understand or it's too much to take in!

I have clear memories of loving the look of our windows in the showroom, then starting to glaze over as the salesperson talked us through the different options. It's so easy to get befuddled as the conversation gets littered with definitions and data. So get your head around the options before you go.

There are a number of different finishes to consider:

- **timber–aluminium:** timber inside and aluminium outside
- **uPVC–aluminium:** uPVC with an aluminium exterior
- **uPVC:** uPVC inside and out.

Consider how your house will look once the windows are in place and walk the streets checking out what other people have done. Old houses can look very different with uPVC windows. But these are often the most economical option. So you need to decide where your priorities lie.

From a climate perspective, think back to digging and which options require materials that have to be dug out of the earth, releasing carbon on the way (see chapter 4). Aluminium is definitely not environmentally friendly. It has to be mined and uses vast amounts of energy in production. BUT it is infinitely recyclable.

I'm never sure of recycling – it's too dependent on people actually making it happen for my liking. So a good question to ask your prospective provider is what they do with the windows they take out – are they actually recycling the glass and the aluminium?

We've got no windows

We waited a long time for our windows. Nothing to be done about it – we retrofitted during Covid – but it did mean that life was b****y cold! We have a 3.5m window and French doors plus one large square window – all of which were gaping holes – and at least one more month to go.

I can't recall now where I heard about it, but someone suggested I try a local window supplier and ask what they had taken out recently. So off I went, fingers crossed, to the nearest window showrooms. I asked, as nicely as I could, whether they had any old windows going spare that we might use in the interim.

I was shown out to the back and found a huge pile of windows and doors just waiting to go to the tip. We rescued a pair of sliding doors that we put end to end to fill the big window space and topped it up with packages of insulation.

The French doors' space was filled with a wooden double door found on Freecycle. It was weird looking and draughty, but significantly better than the alternative.

So if you have gaps waiting for windows go and ask. The worst they can say is no.

WINDOWS FOR FUTURE-PROOFING

Thinking double or triple glazing for wintertime is second nature to us now. But the issue of overheating is only just coming onto our radar. The term here is solar gain - that is the heat created in your home when strong sunshine comes through the windows.

Part O: The problems caused by solar gain and overheating are so significant that a new section has been added to the building regulations. Part O addresses the risks of overheating in new homes, particularly with reference to solar gain.

Although it is not statutory for retrofit, Part O* is worth paying attention to. It's there because it's needed, so why would you ignore it? It's worth checking that your provider and your builder are familiar with Part O – that way, they can help you get the best outcome for every time of year.

Part O covers:

- **solar gain control:** strategies to limit the amount of solar heat entering buildings, such as using shading devices (e.g. external shutters, overhangs or solar control glass)
- **ventilation:** adequate natural or mechanical ventilation systems to remove excess heat from the building (see chapter 10)
- **thermal insulation:** insulation plays a role in reducing heat transfer and can help with overheating (see chapter 11).

If you have a room that gets really hot on sunny days, then you are experiencing solar gain.

Our front room gets the morning sun and it's lovely to sit in there during the cold winter months. In summer it can get a bit hot, but the sun is only there for a short time before setting off over the neighbour's house. We don't see it directly again until sunset in the kitchen, so we've taken no action – just enjoy the bit of solar gain when it's there.

If your house has loads of windows then heat from the sun could be a problem. The risk is overheating in the summer and finding it difficult to cool down, unless you have air conditioning.

It sounds a bit dramatic, but hotter days are coming. We've already experienced days of up to 40°C so we know just how uncomfortable it can be. If we want our homes to be fit for the future, we have to consider heat with every retrofit and renovation project we take on.

The best option is to reduce the amount of sun that comes into the building and glazing is a big part of this.

Managing solar gain

Start by considering whether solar gain is beneficial in your home. Short bursts like we have are pleasant, add a bit more warmth and don't threaten the comfort of the home. But if you have large windows – bifold doors or floor-to-ceiling windows – facing the sun for a good part of the day, then you'll need to reduce the amount of heat coming into the house.

This is an important question to ask any glazing provider. Your first discussions will probably focus on heat retention. But now, as we see climate changing, this conversation needs to include the management of heat. Just don't rely on the salesperson to bring it up.

Think through the impact of the sun on your house:

- which living areas face directly into the sun?
- how long does the sun come directly into your room?
- does the sun cause your house to overheat?
- do you dread sunny days?
- do you rush to draw the curtains before the sun comes up?

It's always a balancing act. It may be that summer sun makes certain areas of your home unbearable, but it's delightful when winter sun is shining. In that case you'd be better looking at double glazing and a cooling system so you can have the best of both worlds.

If the sun is always a problem and you just can't bear the heat, ask about triple glazing – that will keep you warmer in winter and reduce the solar gain in summer.

WHAT YOU NEED TO UNDERSTAND ABOUT GLAZING

Shade and canopies

I visited a friend in her Passive House. Solar gain was an important part of heating the home, so the build has been positioned to get the greatest support from the sun.

However, even a Passive House can have too much of a good thing. While I was there the sun come out in a particularly robust way and I was amazed to see an external canopy begin to roll down, shading the room.

She hadn't moved a finger! It all happened automatically. The canopy/awning was programmed to respond when the sunlight reached a particular intensity, keeping the house cool in summer.

It was incredibly effective and worked whether she was there or not. So no risk of coming home to a furnace.

Shading is an important tool in the struggle with solar gain and worth considering as summer heat increases. Ask your window providers about this too – many include it in their offering.

Glazing

As a rule, choose:

- triple glazing if you want to cut solar gain
- double glazing if you want to keep solar gain.

Providers will certainly tell you the benefits of different glazing in keeping you warm. They will recommend triple glazing on east-facing walls where you get little sun – so cutting out the cold – and double glazing where you can benefit from solar gain.

What you need to do is open out the conversation about heat. Hopefully they'll know how to help you. If not, then try a different supplier or ask them to do some exploring and come back to you.

Don't expect help from your builder on this. Make this decision yourself and tell them of your plans. Visit showrooms, talk about the options and consider what your house needs. Don't let 'the way we've always done things' stop you getting what you want.

FITTING NEW WINDOWS

These guys are experts – right? Why would you need to ask if they know how to fit windows? It just seems rude.

Take it from me, it's not a given and you should be prepared to ask at the outset. This is particularly relevant when you've gone to the trouble of making your home airtight. We assumed expertise and came unstuck, so I would encourage you to ask questions before ordering.

You need the windows to fit well:

- where the glass fits into the frame
- when the window closes
- where the whole window fits into the aperture.

Always check these areas when you've had new windows fitted. It sounds like a no brainer, but there can be mistakes. It may also be that the window or door needs a small adjustment to make it fit better in the frame. Just test with your hand or borrow a thermal camera.

When you've made your house airtight, the windows need to be sealed in with the airtightness tape to prevent air leakage*. They also need to sit in the middle of the insulation to reduce the risk of cold bridging (cold surfaces that will cool the air) and have overlapping air and water barriers.

Get a rough idea of what is needed, then ask about airtightness when ordering your windows. If there is no sign of recognition, try another supplier. You've worked too hard to have this spoil your efficiency.

One final point: new windows and doors have a handle that lifts up to create the seal and so you can lock it. Unless you do that, you could have a small draught coming through. So get into the habit of closing the door and lifting the handle. It's taken me a while to train the family but they've got there in the end!

Summary

- You will be told the U value of the window – a measure of how well the window transmits heat. You don't want it to transmit heat, so you're looking for a low number.
- There are a range of options available, from double glazing to triple glazing and including secondary glazing.
- There are advances in the technology all the time so update yourself before visiting your first showroom.
- Study the options for the finish. Think through what will suit your home.
- Ask questions about solar gain. Windows need to protect us from heat as well as cold. It's unlikely you'll be told about this, so ask the question.
- Ask about experience of fitting windows in airtight houses. Make sure they know what they are doing.

To find the information indicated by the * follow the QR code.

NOTES

CHAPTER 15
WHY RECYLING WON'T OCCUR TO THE BUILDER

Every evening, after the builders had gone home, John would inspect the skip. He found all sorts of stuff that could be recycled or rehomed. Mostly from our work, but sometimes an item the builder brought in from home or another job. I distinctly remember a table lamp that we'd never seen before!

I have to admit, it was irritating to find stuff back in the house. I can understand the builders' need to tidy – even a little bit – and dumping is so much quicker. 'Out of sight, out of mind' works OK for me too – sometimes.

But John is fastidious. He just can't contemplate waste. He pulled all the cardboard out for recycling, rescued bricks that could be rehomed and saved pallets that someone could use for garden furniture.

Sometimes it even meant climbing inside to dig about to make sure he'd got it all. At those moments, I just had to leave the scene. I do have my reputation to think about!

Here's the reality

Every tonne of building material we reuse is a tonne that doesn't need to be dug up, processed, manufactured and transported. Reuse and recycle is one of the simplest, most powerful ways to cut the carbon footprint of your project.

The idea of reusing fits perfectly with retrofit, which is focused on energy efficiency in all its different forms, and waste management is just part of that. Some of the main areas are:

- using water in an economical way
- recycling materials from the build
- using preloved materials in the build.

USING WATER IN AN ECONOMICAL WAY

When you start to think about water, the mind really does boggle. For instance, why do we let our shower water go down the drain and then use freshly treated water, with all the cost incurred in that process, to flush the loo? Why do we allow rainwater to just drain away then use fresh water from the tap to water the garden? Couldn't we use our bathwater to wash the car – or are we really just too dirty?

Unexpected items in the shower!

Working through the edit of this book, I found a lovely comment from Suzanne my editor who said: "A few years ago when my mum was visiting, she got in the shower, then shouted down: 'unexpected items in the shower'. It was the bucket and watering can!"

I know just what she means. John had a large low plastic box that he couldn't bear to get rid of so he brought it up to the bathroom when we were in a

drought. We'd stand in the box, run the shower then transfer the water to a bucket to come downstairs.

Even when we're really mucky and sweaty after a hard days work the plants won't mind. They've got their feet in the earth for heavens sake - they're not proud! So gather your water wherever you may and help your flowers and veggies survive the heat.

We can't do anything with blackwater – that's the posh word for water from the toilet – because it rightly goes straight to the sewer. But greywater management is the way to reduce the water waste we've just accepted until now.

Greywater is anything that has been used before: washing machine water, shower water, washing-up water. All still much too good to just go straight down the drain.

Ways to collect greywater range from the blindingly obvious to the sophisticated:

- **Collect it manually:** place buckets in showers or use a washing-up bowl. This is familiar from hot summers when there's a hosepipe ban and the plants are dying of thirst. Just remember to check the washing-up liquid, washing powder and body products you use - make sure they are natural without chemicals that could harm the garden.
- **Simple plumbing:** install a diverter valve to your washing machine or shower drain to redirect water to a barrel or directly to your garden*.

If you're using greywater for your garden there are a few rules to follow:

- **water the soil, not the leaves:** apply greywater to the base of plants, not on their foliage, to prevent damage
- **target specific plants:** use greywater on established trees, shrubs, hedges and lawns
- **avoid edible plants:** don't use greywater on root crops or leafy vegetables that are eaten raw, unless you know your products are totally non-toxic
- **rotate with fresh water:** use greywater in an area for a while, then switch to fresh or rainwater to allow the soil to flush out any built-up salts or contaminants.

Ultimately, it's best not to store greywater because it can breed bacteria and become a bit smelly. Mind you, any stagnant water can do the same.

Then there are greywater systems

If you want to go further and build in water management, include it at the planning stage. The garden will need to be dug up to access the drains, so if you're building an extension where that will happen anyway, they can be done at the same time.

Systems vary from direct irrigation set-ups that treat water through coarse filters and biological processes to sophisticated chemical-free treatment systems involving filtration and UV disinfection for uses like toilet flushing. Using your greywater provides significant environmental benefits, including water conservation and reduced pressure on wastewater facilities, but means you have to stick to regulations and only use the water in certain ways.

There are different ways of managing and treating greywater that will help cut your bills. The builder is unlikely to talk about this and will probably think you're a bit daft for doing it, but if this makes sense for your home, go searching to find out what the options are.

Questions to ask:

- What chemicals would be used and what impact would they have on the water system overall when the water finally makes its way down the drain?
- What are the practicalities of digging up the garden and how much space will I need to sink a water tank?
- Are there any planning or building regulations that relate to the use and storage of greywater?
- What happens if the system breaks down? Can it be repaired without massive upheaval?

Rainwater harvesting

This will be familiar to most of us – the gutter running directly into the rainwater butt in the garden. When I was a kid, we used this water for washing our hair – it was supposed to make it really soft. It was certainly a kerfuffle, but then washing hair was anyway in the days before showers.

For rainwater harvesting all you need is a water butt – probably not too hard to find on Facebook marketplace or Freegle – placed underneath the down pipe from the gutter. It means you can save the rainwater that falls on your roof and use it to water your plants and vegetables.

The more space you have, the more butts you can have. But be aware that insects can thrive in the wet environment so you need to find some way of covering the butts. There are a number of ways to do this, so go exploring and see what would work in your garden or for your home.

If you're embarking on major work around your house, you could also consider installing a rainwater harvesting tank.

These sit underground so don't affect the aesthetics of your house, but there is a bit more expense and paperwork to deal with. So find out what's involved before making your decision. And include that question: what happens if it goes wrong?

MANAGING YOUR WATER

When considering water, it's worth a quick mention of water softeners and limescale.

If you live in a hard-water area limescale will build up very quickly. This is a problem because it furs up the pipes, making it harder for the water to get through and that means appliances are less efficient. The more you reduce limescale, the less it will cost to do the washing or put on the dishwasher.

Water softeners

Water softeners make your water soft and really nice to use. They also stop the limescale build-up. But to do this they use a lot of salt, which inevitably ends up in the water system.

I live in a very hard water area, so we did have a water softener for a while. There were definite benefits:

- lovely soft water to wash and shower in
- using less product because everything lathers more easily
- no furring up of the kettle or appliances so they last longer and do a better job.

BUT: it soon became a chore. I'd realise the salt needed topping up when I was in the shower, then forget as soon as I got out. I'd forget to buy more salt when I was out in the car. Then there was the problem of storage, not to mention the really thick plastic bags it came in. I know – I need to get a life!

Because it only works with salt in the machine, it is an ongoing expense and I found it a bit frustrating that softened water was used in the loo as well as in the shower. It also means you need a separate drinking water tap. Soft water isn't good for our teeth, as anyone who lives in a naturally soft water area will tell you.

Environmentally it isn't the best. Once the salt has done its job, it goes down the drain and into aquifers (natural reservoirs), harming aquatic life and potentially affecting water quality for we humans.

If you're thinking about having one, do some research to find out if they might work for you. Include the on going cost of salt, the system of delivery and where you could store the bags.

Water conditioners

The alternative is a water conditioner – I was very excited when I discovered this option. It's a small appliance that attaches to the mains inlet so that all incoming water runs through it. It prevents limescale by changing the mineral structure in hard water, causing the lime to form tiny, suspended particles rather than sticking to pipes and appliances. Conditioners don't remove minerals – they alter their structure.

Methods include using electromagnetic fields, which change the shape of mineral particles from tangled to needle-like, or template-assisted crystallisation (TAC).

OK now my mind's going into a pretzel so I'm going to leave you to explore and see what you can find out. I'll put the links I've found on my website, so just follow the QR code at the end of the chapter for a starter.

We definitely need one of these!

In those long days of sitting in a dusty room learning about retrofit and hoping we weren't making too many mistakes, I came across water conditioners. They're another way to manage limescale. No salt needed and no ongoing cost. Buy it, attach it and forget it.

I thought this was a great idea and started calling around to find out where I could buy one. It took a number of calls before I found anyone who knew what I was talking about. So off I went and picked it up. It was a small item that fitted into my handbag and cost about £200.

I've recently discovered that there are different qualities of conditioners available and the more you pay, the longer they last and the more efficient they will be. We definitely noticed a reduction in limescale for the first few years. It was still there, but it came off easily with my fingernail. But after four years it's no longer working so well and the limescale is much harder to remove.

Buying a better quality appliance will ensure the house is covered for 30 years. It not only protects from limescale now, but will also gradually remove the existing limescale. It's on my list to try, so I'll write a blog about it and let you know.

Basic water saving behaviour

In amongst all this technology remember the basic and obvious:

- Take short showers
- Turn off the tap while you're brushing your teeth
- Fix any leaks and dripping taps
- Wash your car with a bucket rather than the hose
- Always run full loads of dishwasher and washing machine

You can also buy efficient shower heads and low flush toilets - so go searching and see what you can find.

REUSING BUILDING MATERIALS

Builders aren't keen on doing this. It can take some work to clean up materials so they can be used again – pulling nails out of wood or cleaning mortar off bricks, for example. The easy way to manage this is to do it yourself – get the material ready for the builder to use when they need it. It's probably cheaper than paying them to do it anyway.

If you live in an old house you could easily reuse some of the original materials:

- picture rails can be cleaned up and used again
- skirting boards can be replaced once floors have been put down
- original doors often look much better
- radiators can be reused
- bricks from a demolished outhouse can be used in a new extension.

If you want to do this, you'll need to be determined. Remember that builders want to do the job, get paid and move on to the next project. Popping to the local building merchant to pick up fresh clean bricks, wood and cement is definitely quicker and easier.

It's heart-breaking

A bungalow was demolished in my road recently. It broke my heart to see the bulldozer ride roughshod over the building. Nothing was saved. An endless stream of lorries blocked the road, taking away load after load of rubble, wood, flooring, windows, radiators, good quality insulation...

It wasn't an old build – probably 1970s – so something would have been of use. Each time I walked by I thought of all the young folk I see on social media doing up their first homes and struggling to make ends meet. There would have been stuff there that they'd really appreciate.

We want life to be easy and we're used to getting what we want quickly, so we forget to reuse. We assume new is better and it feels easier to throw away and start again. But when we put that against the touchstone of climate it hurts. Everything we throw out goes straight to the tip and it will lie there for our grandchildren and their grandchildren to cope with.

So it's definitely worth talking to your builder – you might find they'd be open to reusing. When we lifted our floorboards to insulate under the floor, the builder just assumed we'd get a new lot to put down. Instead we held onto them all. John quite enjoyed the challenge of labelling them all, then doing the jigsaw of putting them down again.

BUYING RECYCLED MATERIALS

This is the other side of the equation. Can you find what you need in preloved form? Now you're looking for someone else's old floorboards, doors or bricks.

Building materials are robust, so old doesn't have to mean ropey. They can be just as durable and hard working – sometimes even more so. We put this learning to good use in our solar barn. We were determined to use as much recycled material as we could. I was amazed at how easy it was. We used:

- wood from the local wood recycling centre*
- a fabulous oak door from a local recycled store*
- a triple-glazed window from the same store
- shelving made from old scaffolding boards
- our old carpet put under the floor to stop foxes burrowing.

It felt so satisfying. A lot more fun than using shiny new stuff. And saved us a lot of money, which is a real help when undertaking a sizeable project.

A gym floor?

I thought John had lost it when he told me he'd found an old gym floor* to go in the barn. We decided against having the gym markings, although in retrospect it might have been fun.

Turns out it was a floor made of solid beech! We never would have afforded that new. Even engineered wood floor, where a thin piece of real wood is stuck to cross-layered plywood, costs a fortune.

One of the measures of a wood floor is how often you could sand it down once it starts to look scruffy. A quality engineered floor will take up to three sandings. A solid wood floor will take loads more than that. This gym floor will see us out.

Our only problem was being a bit cheapskate. The barn is about 2m by 8m and John spent £40 per square metre on the floor. Buying one at £80 per square metre would have made it easier to lay. And it would still have been a remarkable bargain for a product that costs £200 per square metre new.

The advantages of recycled

I recently met someone who told me about the recycled bricks he'd used on his extension. "They fit the house so much better, they look great and they cost a lot less." He was as excited about his bricks as he was about the extension overall!

There is real satisfaction in using materials that have a history. It's also true that many recycled materials will be better quality than the new stuff we find in builders' merchants today.

If you have an older house you may also find recycled or preloved materials work much better with your home. New bricks can look a bit odd when used on a Victorian house – it just makes it harder for an extension to blend in.

So find your favoured recycled/reclaimed/preloved sources and start there before going to the builders' merchant. You can try:

- Facebook marketplace
- Freegle
- wood recycling yards
- architectural salvage
- junk yards
- shops attached to your local tip
- local contacts: WhatsApp groups and networks.

When you talk with your builder, tell them about your ambition to reuse. They may have other projects that are chucking the exact materials you need. We've found the odd floorboard or the few extra bricks we needed by doing this.

If you're buying second-hand materials, always check them over thoroughly. They still need to comply with current safety standards. Windows, for example, should still match glazing safety regulations and you'll need to take a thorough look at structural timber to make sure there is no rot or woodworm.

Cost and climate in one move

Using reclaimed materials is a win both for your bank balance and for the climate. When we reuse, we reduce the digging and that means a reduction in carbon emissions. And when you buy something that was going to be thrown away it is often a lot cheaper – or even free.

Look around to see what you can find. And be generous with any extras you have. Offer them out to others so they can be used fruitfully rather than dumped for future generations to struggle with.

Anyone want some wood fibre?

My friend saw a note on a community site about a homeowner who had over-ordered on wood fibre insulation. He was offering it for free - it couldn't be taken back because of how it had been stored.

She spoke up and went round to the house, only to discover there was enough material to insulate the whole of her house!

This was a great lesson in storing materials. You need a safe, dry area to protect your materials so they're not damaged, but also in case you need to send any back. We didn't get organised in time, so lived for weeks with a dining room full of wood fibre.

The homeowner's loss was my friend's gain. What a generous gift and what a win! She'll dine out on that for years to come and enjoy her warm house even more knowing it was salvaged. And the homeowner who passed it on will know that they did a really good thing.

Summary

- Watch what the builders are throwing out. Make clear that you want to recycle, reuse or pass on what's not needed.
- Look into ways of recycling your greywater – the water that comes from showers, washing machine, dishwasher…
- Gather your rainwater for use in the garden. It saves on your water bill and will be better for the plants.
- Study the pros and cons of a water softener before buying. The water feels great but the salt is damaging to the environment.
- Consider a water conditioner – a one-off purchase that can last for 30 years and reduce your limescale. You just install it and leave it to do its job.
- Reuse whatever you can from your own project. No need to throw good wood or brick just to buy again new.
- Look for preloved materials – they are often cheaper and you can get quality you couldn't afford new. Plus you're holding the carbon in your home rather than leaving it to release as the material decomposes on the dump.

To find the information indicated by the * follow the QR code.

NOTES

CHAPTER 16
WHY HEALTHY HOMES AREN'T ON THE BUILDERS RADAR

Of course we all want a healthy home, but how many of us know what that actually means?

It's sad that homes are not automatically healthy, but the reality is that loads of people live in substandard properties that leave them cold in winter and overheated in summer, with all the illness that goes along with that.

We've already discussed many of the issues here, especially when it comes to building materials, so I'll keep this short and refer you to the chapter that has more detail.

Think about your own home, as well as the homes of people you know, as you consider these questions:

- Are you warm and comfortable even on the coldest days?
- Do you have fresh air in your home?
- Is moisture managed well or do you have condensation and mould to deal with?
- Can you sit in good light and read a book?

- Are you confident you'll be cool on a really hot summer's day?

If you can answer positively to all those questions then you're doing exceptionally well.

The UK Green Building Council* has defined a healthy home, looking at all the different elements that ensure we can live easily and comfortably in our neighbourhood. I'm delighted to see this has taken hold in the building world.

However, I'd be very surprised if your builder speaks to you about this as an overall strategy, although it's exactly what we should be aiming for with any renovation on our homes.

Builders will, of course, know some of the elements of a healthy home. What you need to work out is what they actually know or think they know. Either way, it's time to get in there and work with them to include the elements they haven't yet heard of.

WHAT BUILDERS DON'T KNOW ABOUT HEALTHY HOMES

I'm going to focus on the elements that will be challenging for a regular builder and the elements most likely to be forgotten or left out. So you might say this chapter is your homework!

You're looking for a home with:

- the ability to keep you warm in winter and cool in summer
- no toxins in the air from building and cleaning materials
- a continuous flow of fresh air
- clean drinking water
- manageable energy bills that don't cause stress

- décor that fits what you enjoy and soothes you
- sufficient storage for all your stuff.

Warm in winter, cool in summer

This becomes increasingly important as the climate changes. It's only recently that European homes have had to manage excessive heat. For countries like Spain, France and Italy we're talking extreme heat that leads to forest fires and risks to health. In the UK the heat is not yet so extreme, but still far more than the country is used to.

This means that we need to think about keeping heat out – as well as keeping it in – when improving our homes. Your success in that will depend on using the right materials in your home.

This is definitely an area your builder will struggle with. They'll want to use the familiar synthetic insulation PIR, but there are loads of reasons for not using it in a healthy home. Not least that it will let summer heat into your home in a very short space of time.

Remember decrement delay is the time it takes for the inside of your home to be at the same temperature as outside. It's how easily heat passes through the walls/insulation to affect internal temperature. The relevant times are:

- PIR: 1 to 1.5 hours
- hemp: 9.5 hours +
- wood fibre: 9.9 hours +.

You can see from just these three numbers that if you have PIR insulation your home will struggle to keep you cool when the really hot summer days take hold.

There's plenty more information about this in chapter 11.

Toxin free home

It's hard to take in that we might be poisoning ourselves in our own homes! Yet VOCs are in so many of the products we use, including building materials. There's more detail about this in chapter 6.

Volatile organic compounds (VOCs) are a group of carbon-based chemicals that evaporate at room temperature, releasing gases into the air. They're pollutants that cause a number of health problems such as headaches, ear, nose and throat irritation, and nausea. They can also have long-term effects on kidneys, liver and the nervous system.

Your builder won't know that VOCs are in PIR insulation, glues, fillers, paint, solvents, to name but a few. They might be pleased to understand, so they can take better care of themselves – it's rare to see a builder who wears protective clothing. Or be prepared for them to think you're being fussy, overcautious and even neurotic.

In which case it's for you to be clear about what you want. Chapter 11 tells you all about natural insulation and building materials, none of which will contain damaging VOCs.

Let's talk about gas: while we're on the subject of toxins in the home, I must mention gas. A lot of people are still on gas for their central heating, for cooking and sometimes a gas fire. You need to know that your gas hob gives off particulates while you're cooking. As a result the air quality in your kitchen will be worse than the air quality by the side of a busy road*. So:

- open windows and put the extractor fan on full when you're cooking
- make sure you get good ventilation in your house altogether

- get off gas as soon as you can for your own health.

Continuous fresh air

This is a hugely important topic that is spoken about rarely. We are all so focused on insulation and cutting bills that we forget we need to breathe. There are a number of reasons for caring about fresh air:

- reduce moisture in the home – remember that 20 litres of water produced every day by a family - it has to go somewhere
- avoid the development of mould – a cold home with poor ventilation is the ideal breeding ground for mould spores
- keep energy levels high – without a flow of oxygen we soon get tired and lethargic
- get rid of toxins/VOCs from products and items in the home
- rid the home of the smells of daily life.

You'll find plenty of detail about ventilation in chapter 10, including the options available to you. Just make sure to read and digest. When there are problems with the fabric of the home, moisture and water are very often the culprits.

Without ventilation:

- moist air has nowhere to go so it finds a cold, quiet corner to condense, providing the perfect environment for black mound spores
- visitors know exactly what you had for dinner the day before
- allergens swarm the house, making everyone sneeze, cough or wheeze.

Thinking about air quality also means considering the products you use in your home*. It's disturbing how many VOCs come off the stuff we use to clean the sink, polish the wood, spray the worktop. Not to mention hairspray, liquid soaps, bath products… We live in a soup of toxins of our own making.

This is particularly important if you've made your home airtight, because all those useful draughts have been cut out. Your ventilation will help to clear the air, but using natural products without harmful VOCs is part of the solution.

CLEAN WATER

We all have the right to clean and plentiful water, but in reality this is becoming a rare resource.

Retrofitting is the perfect opportunity to install systems that improve our efficiency with use of water. At present we use the same quality of water for everything, so freshly treated water goes down the loo, through the washing machine or on the garden, just as much as it goes into a glass or cup of tea. That's a pretty crazy way to treat a diminishing resource.

There are options for how to optimise water usage:

- **greywater systems:** take water from the washing machine, shower and sink, treat it and reuse for flushing the toilet and watering the garden
- **rainwater harvesting:** gather rainwater to use in the garden
- **low-flow fixtures:** choose fixtures that reduce the amount of water used – e.g. shower heads, taps etc.

All these options can be installed during retrofit – it's the perfect time when the ground is already dug up outside.

I wish we'd known about this

There are a number of possibilities that passed us by when we were doing our retrofit and this is one of them. It was the perfect example of not knowing what we didn't know. If we'd seen anything about it, we'd have been searching and asking every expert we could to discover our options.

To do it properly after the retrofit is a big deal. Using greywater to flush the loo would mean digging up the drains and I'm not sure I can face that again so soon. However, harvesting rainwater is a much more straightforward job.

So far we have a water butt collecting all the run-off from our garden barn. We have a plan for next summer to take water from the shower for the garden. I say 'we'. John is the gardener, so he has his mind around this already. We have a wonderful, lush garden, perfect for pollinators, growing veg and supporting wildlife, so keeping it moist and growing well is really important.

You may feel that a retrofit is enough on its own and you just have too much to think about. But if you can find some time to consider water management it's a very good choice for future-proofing.

Look at chapter 15 to find out what you can do to conserve water and get best use from it.

Manageable energy bills

This is the ultimate target of retrofit – that you can be safe and comfortable in your home without being afraid of the energy bill coming through the door. In reality this is what the whole book is about.

It means:

- knowing what to ask your builder for so you get an energy-efficient home (chapter 3)
- insulating, making airtight and ventilating your home so you use the least energy possible (chapter 10)
- using sustainable heating – heat pumps are the most likely option – so your heating is truly energy efficient and you don't need to use fossil fuels (chapter 12)
- choosing appliances that are the most efficient you can find so the meter doesn't shift when you put them on (see *Beginner's Guide to Eco Renovation*)
- installing renewable energy to fuel your own needs and to sell back to the grid (chapter 13).

Decor and storage

This is not an area covered in this book. But there is a great deal out there for you to find on the web.

While décor and storage can feel a bit minor in the face of climate change, it is extremely important to our sense of well-being. Feeling comfortable in our home and proud to invite people in is top of the list when all the basics are in place.

The good thing about décor is that we can add this in gradually as budget allows. We can allow our style to emerge, change what we no longer enjoy, go on the hunt for exactly the right colour or texture.

It is also a great opportunity to explore the world of preloved and vintage home décor. Finding the piece of furniture you want second-hand and bringing history into your home is a great way to decorate. It can add meaning as well as comfort and become a great talking point.

Nothing uncomfortable

The renovation before the recent retrofit was a shift into our GranPad – the ageing version of a bachelor pad. We had a house full of furniture that had brought up two kids, three dogs, two rabbits, a guinea pig and a cat. It was all much loved but very well worn.

We gutted downstairs and went for new furniture throughout (hadn't entered the world of preloved all those years ago) . It was a wonderful experience – we even got help from an interior designer who stiffened our backbones and encouraged us to be bold.

My main concern was comfort all the way. Of course I wanted beautiful. But I also wanted somewhere to stretch out and slouch, somewhere to sit comfortably when my back was struggling, and to feel at ease when sitting at the dining table. I wanted to feel proud and have people admire my home. And most important, I wanted my grandkids to relax and enjoy themselves without me always having to say no/stop/don't jump.

The end result? We have a home that is totally suited to our needs. Every piece of furniture is comfortable and/or beautiful according to our needs and standards. Everything is also durable – and aren't I pleased we stuck with that! My youngest grandson is growing like a weed, but still loves to jump on the sofa and spread Lego all over the floor.

You know you've got it right by the sigh of contentment when you come through the front door. Home is comfy, safe and robust – all at the same time!

Décor is very personal and we all need to do it our own way. Climate-wise, the more we can find preloved the better the world will be. Less carbon released in manufacture of new products, less spent in travel from far-flung locations, less VOCs to be released into the home.

Summary

- You need your home to be warm in winter and cool in summer – something you can achieve through the use of breathable materials.
- Cut down on your use of toxic materials – both in the building process and in your day-to-day beauty and cleaning products. The VOCs they give off are polluting the air in your home.
- Make sure you have adequate ventilation. This is missing in the majority of homes and it leads to condensation, mould and air pollution.
- When renovating or retrofitting, use the opportunity to install greywater systems. Reusing shower or bath water cuts down on water bills.
- Use every opportunity to make your home energy efficient. It will reduce your energy bills and your carbon footprint.

WHAT THE BUILDER WON'T TELL YOU

To find the information indicated by the * follow the QR code:

NOTES

CHAPTER 17
INSTAGRAM WISDOM

I see so many people on Instagram working on their forever homes. Some of the projects they take on are mind blowing - I would never have been so brave in my early years!

While writing this book, I kept all those folk in mind:

- what did they need to know?
- how could I help them make the perfect home for the future?
- what might make their lives easier

So I put up a post and asked a question. Remember, responses can come from anywhere in the world, so we get an insight into the challenges faced in other countries.

WHAT DO YOU WISH THE BUILDER HAD TOLD YOU?

This is what they told me:

The builder won't tell you that he hasn't built this way before so he might need lots of support from you. **@enbeearchitecture**

That what they are installing may be toxic and bad for the environment, and you COULD have a choice for healthier and planet forward options, IF they let you choose. **@idgreenlist**

Get to know your own building and how it works through the seasons before making significant decisions - especially with fixing issues **@Expertible**

The builder won't tell you that he hasn't built this way before so he might need to charge you more when he discovers how complex it is. **@enbeearchitecture**

That they need to be supervised AT ALL TIMES if you have no project manager. Learned the hard way when I left the contractor to work half a day and the spec we discussed beforehand was not followed.... Also materials going missing from site, massive delay due to the contractor "having another project". **@szh_ng**

They won't tell you what to order to get it on site in time. The best retrofit materials often take a little longer to arrive so don't let poor planning push you towards whatever's quickest from the usual merchant. The right choice is worth the wait. **@herownspace**

If you like a warm house and don't have a ventilation system installed, all your shoes and handbags will get mouldy. Also, your boiler works fine but if you don't have your heating system flushed, your system will be wildly inefficient, your house will be cold and your bills will be 4x higher. **@idgreenlist**

The true cost (which I bet they knew). But also, I have often found that it is not so much what they have not told me, but rather the untruths they have told me (like you must use celotex gold, etc). **@danishstorm**

MOLD is my biggest concern by far!! My family has struggled with mold toxicity and mold sickness for years living in the pacific northwest of the US. It's impossible to escape here as renters but our goal is to buy land and build a mold proof home in the next few years. I'm so happy to have found your page!! **@annacortez277**

Your builder's quoting time is limited, mainly because it's a non profitable task. Therefore they won't usually price for different options. Instead they will ask which finishing materials you want to use and price for that. Fitting costs/labour can become a parameter you loose control of. When working with a budget, it's always worth asking for advice about fitting costs for the options you have in mind. This might help you choose and save you lots of back and forth over pricing. **@theplace_between**

Australia. Let's put it this way, not even double glazing is common here yet. We simply don't have the materials here for much else. I've tried to find the types of insulation you recommend, and plasters like Diathonite, but it's hopeless. The only builders who know anything about eco-building are those working on Passivhaus projects. When you have found something, the builders just say they can't work with it because they don't have the experience with the product and it would make my minor renovations a similar cost to an entire rebuild! **@sirpatrickspens**

One bit of advice that I wish the plasterer had given to our sellers was that to plaster over woodchip wallpaper was not a good idea, especially later when the old built in wardrobes fall apart the "new" plaster falls off the wall. This has been found to be done to every wall in our house and I wondered why the reveals around the door frames were set back!! **@elmwoodsview**

That even brand new builds can have high levels of mould because of the timbers or frame getting wet during construction. **@samanthajosephinewise**

Everything water damage, mould and mycotoxins. That mould remediation experts say the mould is "dead" because it's dry and only remediate what's wet. That nearly every bathroom demolition reveals mould. That the toxic compounds are used in so many building materials through various decades. Depending on your climate, we build differently and this affects air circulation and the buildings breathability. That city water pressure can be too high for your pipes and it's your responsibility to have a pressure control for the water coming into your house. I could go on 😅 **@investederica**

Once you get the breathability and air circulation concept of old buildings it changes the goal posts. Also realise that many do not actually get the concept that if it gets interrupted that opens up a can of worms. It's ok not to know. We eventually found an amazing heritage specialist / surveyor who recommended a national trust stone mason. Gosh they were the dream team and taught us so much!! We would have saved a lot of money if we got the right gents involved early on **@heathertstuart**

Our learnings: 1. Don't trust a builder. 2. Make building control your best friend/keep them close - and verify your builder's say so. 3. Get electricians in first. 4. Don't trust builders with work where there are specialisms for it (sanding floors). Source materials yourself (they'll choose the cheapest most uneconomical and least environmentally friendly). 5. Question everything. Including where recycling waste ends up. **@christinembarkey**

That the pipe end for the shower doesn't go directly below where you want the drain hole of the shower to be. And that grout isn't waterproof (that last one blew my mind when I found out) **@fransflat**

Based on very recent conversations with my builder:

1. You don't have to have a huge, expensive retrofit of your whole house for a heat pump to function (though insulation definitely helps).

2. That underfloor heating is often more efficient than radiators and it can be retrofitted without taking up floorboards.

3. There are ventilation options outside of just opening windows.

4. Different types of window (timber, UPVC, aluminium) have different levels of insulation.

5. Different types of plaster and flooring can affect circulation and temperature in the house.

6. There's a huge market online for used doors and windows, so you don't have to throw them in the tip! (That last one is still a sore point!) **@the_budget_bungalow**

Wood windows are economical & better than pvc/plastic windows. They fit your home perfectly because they are individually made FOR your home. People who think they are draughty don't know about window glazing & winter storm windows. You just swap them out for screens in summer, glaze the windows as needed, & they are amazing. The pvc and plastic windows grow mould incredibly fast and are awful **@ethereal_mews**

That plaster should be saved!! Old lath and plaster homes are way more efficient than drywall. And bonus…plaster doesn't go mouldy!! **@serenity_hill_farmstead**

Be prepared to source materials yourself. Once I had a log house and exposed a wall of the staircase after a new support wall was installed. I located a log to replace an original on a wall that was termite destroyed. After crafting the joints for a good fit; we removed the "new log" and proceeded to test for termites. I wrapped it in plastic and placed it on a large iron griddle in the sun. In 2 weeks a mass of termites appeared under the plastic. I waited another few weeks, called in an expert and had it treated. The log was a beautiful match and retained the original look of the grand colonial house.
@wattjoni

Plastic roof fascia hides a lot of issues. Rotten roofing lathes, needing new roof felt and lathes replaced. £10K later we had a leak proof roof. (none of it picked up by the surveyor)
@elmswoodview

GLOSSARY

ACH – air changes per hour. Measure of airtightness.

ASHP - Air Source Heat Pump.

Air bricks – Regular household bricks with holes in them for ventilation. Used primarily to create air flow underneath a suspended ground level timber floor to reduce condensation. These must never be blocked up or covered.

Airtightness - removing draughts (uncontrolled air) from your home by adding an breathable airtight membrane.

Block and beam floor – floor made of suspended pre-cast concrete beams and blocks. The beam spans the space between the supporting walls with a lightweight concrete block as an infill.

Breathability – The transfer of moisture through a material or building element.

Breathable paint – non toxic paint that is fully breathable with negligible VOC's, plastics or heavy metals. The final layer of a breathable wall.

GLOSSARY

Carbon sink - a natural or artificial reservoir that absorbs more carbon from the atmosphere than it releases.

Clay plaster - a breathable, non-toxic, and eco-friendly wall finish made from a mixture of natural clays, sand, and plant fibres.

Compacted hardcore/6F2 – crushed recycled concrete bricks and mortar for use as hardcore in floor construction.

COP – Coefficient of Performance. A measure of energy in compared to energy out at a given moment. A high COP of 1:4: one unit of energy and get four units of energy out. 1:1 means you get just one until of energy out for every unit put in.

Decrement delay – the ability of insulation to hold heat and slow the time taken for the outside temperature to be replicated inside.

Density - refers to the mass of a material per unit volume. The measure of a material's capacity to absorb, store and release heat over time.

Diathonite – thermal plaster that can vary in thickness according to need. Diathonite can be used both internally and externally.

Embodied carbon - refers to the greenhouse gas emissions released through the entire life cycle of a product.

EnerPHit – set of standards to be achieved in an existing house so it has a high energy efficiency and reduced energy demands.

EPC – Energy Performance Certificate. Measurement of how energy efficient a home or a product is.

Estimate - a non-binding 'guesstimate' of potential costs.

GLOSSARY

EWI – external wall insulation.

Foam glass aggregate – alternative to concrete floors. Made from recycled glass.

Future proof - anticipating future challenges and changes and taking steps to minimise their potential negative impacts in your life and home.

Heat recovery ventilation - a system that provides a continuous supply of fresh, filtered air while recovering the energy from outgoing, stale air. Can be single room or whole house system. MVHR - Mechanical ventilation with heat recovery.

Hemp – an eco-friendly and sustainable building material made from the strong, woody fibres of the industrial hemp plant. Excellent for staying warm and cool.

Hempcrete - Hempcrete is a lightweight, bio-composite building material made from mixing hemp shivs (the woody inner core of the industrial hemp plant) with a lime-based binder and water.

IWI – internal wall insulation.

Lime plaster – a traditional plaster used in old and eco buildings. Made of a mixture of lime, sand and water, lime plaster is breathable and eco friendly.

Microbore pipes - pipework 10–12mm in diameter (about the size of your little finger). Regular pipes are 15mm or larger (about the size of your middle finger.)

MVHR – a whole-house ventilation system that runs constantly in the background, holding warmth through a heat recovery system.

Off gassing – chemicals / pollutants that are released from petrochemical plastics in your home.

GLOSSARY

Passive house - a building standard that prioritises extreme energy efficiency to achieve comfortable indoor temperatures with very little energy for heating and cooling.

PIR/PUR – rigid insulating material made using petrochemicals with high embodied carbon. Brands include Kingspan and Celotex.

Principle designer - the person responsible for health and safety issues on site through to the safe delivery of the project according to building regulations.

PV – Photo voltaic solar panels that convert light into electric power.

Quote - a legally binding calculation of the actual costs.

R value or thermal resistance – how a material resists the flow of heat. Higher the better.

Renovation – to modify an existing house, or build a new part of the house from scratch: new extensions, opening up the loft in a house etc.

Retrofit – to improve how a home functions. To upgrade an existing building adding in systems that weren't there previously – insulation, airtightness, ventilation, solar, renewable heating.

SCOP - seasonal coefficient of performance, describing the average over time. See COP above.

Sisal - a fibrous plant grown in warm, moist tropical climates. Mixed with other fibres like wool and sacking it becomes an excellent thermal and acoustic insulation.

Solar gain – the amount of heat that your windows allow into the room. Manage solar gain in order to not to overheat in summer.

GLOSSARY

STROCKS™ - structural block of clay-rich earth and chopped straw. Used as the inner skin of an external wall and for internal load-bearing walls, typically up to 3 storeys.

Sub floor void – the gap between the earth and the joists under a suspended floor. Important for air bricks to be unimpeded in the void to avoid condensation and mould.

Suspended wood floor – a floor that is supported on timber joists, with a void below.

Thermal bridge, cold bridge - an object or area that hasn't been fully insulated. Means it loses heat, increasing the risk of condensing, leading to mould.

Thermal conductivity/K value/Lamba value - how heat travels through a material by conduction. Lower the better.

Thermal diffusivity – how a material holds onto heat, then releases it. Made up of: high specific heat capacity + low thermal conductivity + high density.

Thermal envelope - The thermal envelope is the barrier around your house separating inside from outside – the walls, floors, roof and windows all contribute to that barrier. There is always a thermal envelope, it's just not always very efficient.

Thermal Imaging camera – a handheld device that detects infrared energy (heat) and converts it into an image. Thermal cameras are used for troubleshooting eg: identifying cold spots, poorly fitting windows, inadequate insulation……..

U Value – measures the heat lost through a given material, or group of materials that form a building element (such as a wall, roof or floor). Used to describe efficiency of insulation materials and double/triple glazing etc.

UFH - Underfloor heating.

GLOSSARY

VOC – volatile organic compounds. Invisible gases that build up from cosmetics, air fresheners, hair spray, cleaning fluids and off gassing, the chemicals released from plastics in your home.

Water conditioner - small appliance that attaches to the mains inlet; all incoming water runs through it. Prevents limescale.

Wood fibre insulation - robust and flexible insulation that can be used for floors, roofs and walls. Has low U values and is 100% compostable and recyclable at the end of its life.

Zero Disrupt – Heat Geek installation of heat pumps at less than the cost of a gas boiler.

INDEX

A

Air leakage, 25, 162, 265
Air-permeability test, 161
Air pollution, 20, 47–48, 291
Air source heat pump (ASHP), 6, 19, 38, 211–212, 226–227, 229, 231–232
Airtight membrane, 37, 62, 160, 163, 199
Airtightness:
  Blow test, 161–164
  Breathability, 32–33, 96–98, 154, 179
  Draughts, 158–159, 164, 257
  Energy efficiency, 22, 37, 79, 268
  Sealant, 51
Argon gas, 255

B

Battery, 235–236, 240–241
  Back-up power, 241, 250
  Bi-directional charging, 243
  Environmental impact, 51, 241
Beginner's Guide to Eco Renovation, 10, 16, 129, 158, 173, 289
Bi-directional charging, 243
Bifacial PV panels, 247
The big five, 153
  Airtightness, 153–154, 158, 160–161
  Renewable energy, 154, 234
  Sustainable heating, 154, 210
Bio-SIP, 191
Blow test, 161–164

INDEX

Builder costs:
   Estimate, 102, 106, 110, 122, 140
   Final invoice, 121
   Quote, 59, 100, 102, 106, 108–109
   Timely payment, 117
Building professionals, 127
   Architect, 128–129
   Architectural designer, 128–129
   Building inspector, 128, 132–133
   Interior designer, 128, 130–131
   Retrofit coordinator, 128, 134–135, 137
   Structural engineer, 128, 131–132
Building regulations, 127, 136, 138–139, 171
   Building Control Completion Certificates, 136, 148
   Certificate of lawful development, 136
   Local authority, 28, 133, 136, 237
   Private building inspectors, 133
   Safety Regulator, 133
Building, working styles:
   Design and build, 106, 112, 114, 118, 122
   DIY, 106, 115–116, 158, 167
   Individual tradespeople, 106
   Project management, 107, 134–135

C

Capillary active, 93
Carbon emissions, 4, 44, 188, 213, 232, 279
Carbon monoxide, 90
Carbon taxes, 219
CarbonLite, 40
Cardboard, 82, 199, 267
Cellulose, 51, 91, 191, 197, 199
Cement, 48, 71, 86, 95, 156, 188, 275
Clay plaster, 17, 180, 200, 205–207
Closed-cell, synthetic insulation, 187
Coal, 19, 45–46, 48, 53, 191
Coefficient of performance (COP), 218–219
Communication, 61, 103, 105, 215, 229
Compacted hardcore/6F2, 190
Concrete, 48, 51, 86, 156
   Modern brick, 87
Condensation, 20, 27–28, 32, 50, 138, 157, 257
Conduction, 81

Continuing professional development (CPD), 129
Contract, basic clauses, 145
    Accidental damage, 148, 150
    Accidents, 148, 151
    Agreement about managing, 148
    Change of design, 147
    Consideration, 4, 145–146, 243
    Dates, 145–146
    Dispute resolution, 146
    Extras, 105, 147, 279
    Liability insurance, 147
    Payment plan, 146
    Retention clauses, 148
    Termination, 146
    Waste management, 147, 268
Contracts, elements of, 145
    Acceptance, 145
    Consideration, 145–146
    Intention, 145
    Offer, 145
Contracts, practical help, 149
    Federation of Master Builders, 59, 150
    Joint Contracts Tribunal (JCT), 149
    The Place Between, 150
C.U.In glass, 258

D

Damp/wet wall, 32, 41, 155
Decrement delay, 85, 168–169
Dew point, 169
Diffusion of Innovations, 215
    Cold bridging, 265
    Early adopters, 215–216
Double glazing:
    New, 257
    Old, 257
Draughts, where to find, 164
    Ceiling lights, 164
    Doors and windows, 164
    Down chimneys, 164
    Floorboards, 159, 164
    Pipe inlets, 164
    Skirting board, 164

E

Eco-architect, 6
Eco builder, 6, 23, 51, 79, 158, 185, 234
Eco-friendly, 54, 62, 89, 96, 169, 189–190, 195, 201
Eco homes, 21
Eco Renovation, 10, 16, 129, 158, 173, 289
ECO4 grant scheme, 252
Electric vehicle (EV), 53, 243
Embodied carbon, 76, 78–79, 89–93
    Construction, 47, 78
    End of life, 78
    Maintenance, 78
    Manufacture, 78

INDEX

Raw material extraction, 78
Transportation, 78–79
Energy cost, 55, 246
   Marginal wholesale price, 218
Energy-efficient homes, 1, 37, 40, 157
Energy Performance Certificate (EPC), 9, 25, 37, 135
EnerPHit-level retrofit, 38
European environmental label, 203
   Clean and efficient production, 203
   Protection of the environment and people's health, 203
Everything Electric Show, 227
Expanded and extruded polystyrene (XPS), 48, 87, 91, 157, 191
Extractor fans, 171, 173

F

Federation of Master Builders, 59, 150
Fibre glass, 87, 90
Flemish bond, 196
Floorboards, 18, 25, 28, 154, 158–159, 164, 225, 276
Foam glass aggregate (FGA), 191, 195
Foil-backed petrochemical board, 79
Fossil fuel, 216, 289
Future-proof, 5, 18, 50, 54, 156, 163, 227

G

Glazing, 11, 31–32, 255–265, 279
   Argon gas, 255
   Double glazing, 31–32, 255, 257, 262–265
   Krypton gas, 255
   Triple glazing, 255–258, 260
Global warming potential (GWP), 91, 94
Goodall, Jane, 242
Grand Designs, 78, 201
Graphite polystyrene (GPS), 191
Green Seal, 97
Greenhouse gas, 45, 47–48, 78
Greywater, 269–271, 280, 287
Grid management:
   Bi-directional charging, 243
   Distribution network operator (DNO), 249–250, 254
Ground source heat pump (GSHP), 211–212

H

Healey, Julia, 158
Heat Geek, 218, 220, 222
Heat pumps, 210–221
    Air source heat pump (ASHP), 211–212
    Coefficient of performance (COP), 218–219
    Ground source heat pump (GSHP), 211–212
    Seasonal coefficient of performance (SCOP), 218–220
    Underfloor heating (UFH), 224–227, 231
    Weather compensation, 230–231
Heating alternatives, 212
    Biomass boilers, 212
    Electric combi boilers, 212
    Geothermal systems, 212–213
Hemp insulation, 18, 54, 137, 195, 198, 224
Hempcrete, 51, 137, 187, 191–192
Heritage pane, 258
Homes, Barratt, 14
    Build net zero homes from 2030, 14
House value, 37
Humidity, 22–23, 176, 193, 206
Humidity monitor, 206–207

Hydrofluoroolefins (HFOs), 91, 94

I

Indi-Nature, 198
Insulation, 165–170
    External wall insulation (EWI), 166–167, 186, 202
    Internal wall insulation (IWI), 166–167, 214
Insurance, 94, 100, 129, 147–148, 150
    Employer's liability (EL), 150
    Public liability (PL), 150
Interior decorator, 130
Internal designer, 112
Internal temperature, 83, 85, 284

J

Joint Contracts Tribunal (JCT), 149

K

K value, 81–82, 84, 197
The Kelvin scale, 86
Kilowatt hours (kWh), 36, 252
Krypton gas, 255

L

Lime plaster, 16, 98, 156, 195, 199, 204–205

INDEX

Limescale, 272–274, 280
Low lambda value, 80

M

Materials, basic, 86
  Cement, 48, 86, 156
  Concrete, 48, 51, 86, 156
  Modern brick, 87
Materials, building with natural, 187
  Bio-SIP, 191
  Clay bricks, 187, 193, 195
  Glass aggregate/glass beads, 187
  Hempcrete, 51, 137, 187, 191–192
  Screw piles, 51, 187, 189, 194
  Straw bale houses, 187
  Tyrecrete, 187, 192
Materials, conventional:
  Concrete, 27, 48, 51–52, 86, 156, 161, 184, 188–192, 194, 239
  Foam insulation, 91
  Synthetic material, 8, 10, 50, 54, 180, 183, 197
Materials, high-carbon, 48
  Cement and concrete, 48
  New bricks, 48, 278
  Steel and aluminium, 48
Materials, highly efficient and effective, 51
  Innovative concrete alternatives, 51
  Low-carbon/bio-based materials, 51
  Low-VOC paints and finishes, 51
  Natural insulation, 51, 201, 205, 285
  Recycled and reclaimed materials, 51
Materials, insulation, 87, 136
  Different working models when employing builders, 106
  Fibre glass/glass wool, 87, 90
  Mineral wool, 87, 92
  PIR insulation/PUR, 87
  Spray foam, 87, 93–94
  XPS/EPS, 87
Materials, natural insulation, 197
  Cellulose, 197, 199
  Denim, 30, 197, 200–201, 208
  Diathonite, 197, 202, 208
  Hemp fibre, 197, 205
  Sheep's wool, 197, 201
  Sisal, 197, 200, 208
  Wood fibre, 197–199
Materials, toxic, non-renewable, 48
  Composite wood products, 48, 78
  Fibreglass, 48

INDEX

PVC (Polyvinyl chloride), 48
Synthetic insulation, 30, 48, 88, 284
    Fibre glass/glass wool, 87, 90
    Mineral wool, 87, 92
    PIR, 7–8, 48, 87–88, 90
    Spray foam, 87, 93–94
    XPS/EPS, 87
Mechanical ventilation system, 172, 261
Microbore pipes, 222–223
Microinverters, 238–239
Mineral wool, 87, 92
Moisture, 20, 28–29, 31–32, 186–187
Moisture risk, 40
Mould, 77, 157, 167, 257

N

New Homes Net Zero Transition Plan, 14
    Pathway and metrics for decarbonisation, 14
Noise pollution, 47

O

Octopus Energy, 14
    Zero bills homes, 14
Off-gassing, 30, 50, 77–78, 89–90, 93, 97, 169
Open-cell structure, insulation, 187
Organisations, 149
    Architects Registration Board (ARB), 129
    Association for Environment Conscious Building (AECB), 40
    Distribution network operator (DNO), 249–250, 254
    Ecological Building Systems (EBS), 73
    Energy regulator (Ofgem), 184, 250
    Federation of Master Builders (FMB), 59, 150
    National Federation of Builders (NFB), 59
    Registered building control approver (RBCA), 133
Oriented strand boards (OSD), 191
Overheating: Part O, 260–261
    Solar gain control, 261
    Thermal insulation, 261
    Ventilation, 261

P

Parking, 69–70, 74
PAS 2035 (Publicly Available Specification 2035), 28–29, 34, 40, 134
Passive House, 14, 38–39, 140, 163, 239, 263

Pavatextil, 201
Personal protective equipment (PPE), 89, 116
Photovoltaic (PV), 238, 245–248
PIR insulation, 7, 87, 137, 180, 284–285
Planning permission, 106–108, 128, 136
Plasma-ion treatment, 201
Plaster options, 204
    Clay plaster, 205–208
    Gypsum plaster, 204
    Lime plaster, 204–205
Plasterboard, 95–96, 155–156
Plastic pipework, 222–223
Plumber, 68, 74, 112, 148–149, 164
Polystyrene, 48, 91, 157, 186, 191
Preloved materials, 268, 278, 280
Principal designer (PD), 107, 127, 140
Professional indemnity insurance, 129, 150
Project manager, 108, 112, 116, 133–134
PVC (Polyvinyl chloride), 48

Q

Quality assurance:
    European environmental label, 203
    Natureplus Eco label, 203
    Sustainability of resources, 203
    Trustmark, 60

R

R value, 82, 168
Rainwater, 268, 270–271, 280, 287
Recycled materials, 54, 195, 276, 278
Reducing landfill, 11, 53
Renovation, 5, 16–19, 125–126
Retrofit, 28–31, 41, 153
    Ventilation, controlled, 5
Retrofit advice:
    Expertible, 135, 141
    Her Own Space, 141
    The Place Between, 150
Retrofit assessor, 134
Reusing waste, 11
Rigid solar panels, 246
Roof-integrated PV, 247

S

Scaffolding, 119, 166, 277
Scope of works, 117
Screw piles, 51, 187, 189, 194
Seasonal coefficient of performance (SCOP), 218–220
Secondary glazing, 257, 265
Sheeps wool insulation, 197, 201

Borax, 201
Natural material, 136, 169
Shock ventilation, 172
Signalling intent, 144
Single glazing, 257
Single-room heat recovery units, 37, 171, 173, 175–176
SIPS (Structural insulated panels), 190–191
Skip, cost, 118–119
Solar:
    Batteries, 11, 242–243, 252
    Bifacial PV panels, 247
    Photovoltaic (PV), 238, 245–248
    Rigid panels, 246–247
    Selling to the grid, 11
    Thermal solar, 245
Solar panels, types of, 245
    Photovoltaic (PV), 238, 245–248, 253
    Thermal solar, 245
Solar thermal panels, 245
Solar tiles, 248, 253
Sole builder, 111–112, 118–119
Specific heat capacity, 83–84
Spray foam, 87, 93–94
Stay Warm for Less, 160
Steam diffusion (SD), 97–98, 208
    Breathability, 96–98
    Paint, 50–51, 96–98, 207–208

Trapped moisture, 169
Vapour open materials, 179
Stretcher bond, 27
String inverter, 238
Strocks, 193
Sunamp heat battery, 222
Sustainable heating, options, 11, 33, 289
    Air source, 11, 38, 211–212, 227, 232
    Biomass, 11, 212–213
    Electric combi, 212–213
    Geothermal, 212–213
    Ground source, 11, 211–213
Sustainable insulation, 7, 170

T

Thermal diffusivity, 83–84
Thermal energy, 83
Thermal envelope, 24–25, 160
Thermal imaging camera, 31, 159
Thermal mass, 84, 139, 169, 181
Thermal resistance, 82, 93, 168, 192
Thermal solar, 245
    Thin film panels, 249
Thin film panels, 249
Throat irritation, 77, 89, 93, 285

Touchstone, 7–8, 256, 276
Toxins in beautifying the home, 50
  Furniture, 50
  Toxic paints, 50
  Vinyl and laminate flooring, 50
Trickle vents, 162, 171–172
TrustMark, 60

U

U value, 81, 100, 137–140
UK Green Building Council, 283
Underfloor heating (UFH), 68, 74, 214, 224–227, 231
Unity lime, 192, 202

V

Vapour permeable, 30–31, 89, 96–97, 179, 205
VAT, on retrofit, 106
Ventilation, 162, 164–165, 170–172, 174–175, 177
  Air changes per hour (ACH), 161–163, 171
  Heat recovery units, 16, 37, 171, 173, 175–176
  Mechanical ventilation heat recovery (MVHR), 32, 177–179
  Trickle vents, 162, 171–172
  Victorian sub-floor void, 27

Vinyl, 50, 98
Visit a Heat Pump, 211, 216
VOCs, 30, 48, 50–51, 76–78, 89–93, 285
  In building materials, 77

W

Warm Homes: Social Housing Fund, 28
Waste production, 47, 49
Water:
  Conditioners, 273–274
  Consumption, 11
  Pollution, 47
  Quality, 11, 273
  Softeners, 272
Water management, 270, 288
  Collect it manually, 269
  Greywater, 269–271, 280, 287
  Simple plumbing, 269
Water usage optimization, 287
  Greywater systems, 270, 287, 291
  Low-flow fixtures, 287
  Rainwater harvesting, 271, 287
Weather compensation, 230–231
Window frames, 48
  Timber-aluminium, 259
  Timber frame, 190, 194
  UPVC, 259

UPVC-aluminium, 259
Wood fibre insulation, 140, 199, 279
Wood wool board, 194

Y

York Council, 14
    Building 400 zero carbon homes, 14

ABOUT THE AUTHOR

I've worried about climate for as long as I can remember. Acid rain, a hole in the ozone layer, nuclear waste….. you name it, I've been concerned. And don't get me onto the endless battle with plastic.

The reason for my concern has always been family – my own and yours. I have two fantastic daughters, three amazing grandkids, two cuddly gran-dogs, plus a remarkable husband of 40+ years. We've been through life's challenges in glorious technicolour, but always done it together, learning loads along the way.

At the same time, I've gone through a few careers: teaching, social work, psychotherapist, founding/leading a psychotherapy training centre, executive coach/facilitator and, of course, author. Apart from the first two, always independent and running my own business, which meant that if I needed a change, I had to create it.

So when it came to transforming our home: I needed a book that didn't exist, so I wrote it. I didn't understand retrofit, so I learned. I didn't know others who were as concerned, so I found them. It's been the best journey, shifting my focus from leadership development to insulation – not to mention a lot of fun!

No surprise that I've put all these skills together in my retirement, using them to care for the family and home – mine, yours and the planet – our ultimate home. When I finally pop my clogs, I just want to know I've done my best to leave you all with a safer world to live in.

As a self proclaimed building bore, I love talking about and hearing about retrofit stories, so do let me know how you're getting on. Let's work together to get the word out and make a real difference for all those little people to come.

www.ecorenovationhome.com

Social media: @ecorenovationhome

ALSO BY JUDITH LEARY-JOYCE

This is the book we needed during the eco renovation of our Victorian end of terrace. It didn't exist for us, so I wrote it for you.

This is not a 'How to…' book. It's an overview to help you understand the basics, get your head around the main issues and define the questions to ask your providers. It's easy to read, full of stories from our experience, including the mistakes we made so you don't have to.

If you have your own home and you're going to renovate or retrofit anyway, why not be sustainable and eco friendly, reducing your energy bills and taking positive climate action at the same time?

Follow Judith on Social Media*

Stay up to date with Judith's blog* Eco Renovation Home

www.ingramcontent.com/pod-product-compliance
Lightning Source LLC
Chambersburg PA
CBHW071149070526
44584CB00019B/2715